# GLORIOUS CHINA

www.royalcollins.com

# GLORIOUS CHINA

## The Scientific and Technological Powerhouse Dream

By the Writing Group of *Glorious China*

*Books Beyond Boundaries*

ROYAL COLLINS

Glorious China: The Sceintific and Technological Powerhouse Dream

By the Writing Group of *Glorious China*
Translated by Daniel McRyan

First published in 2023 by Royal Collins Publishing Group Inc.
Groupe Publication Royal Collins Inc.
BKM Royalcollins Publishers Private Limited

Headquarters: 550-555 boul. René-Lévesque O Montréal (Québec) H2Z1B1 Canada
India office: 805 Hemkunt House, 8th Floor, Rajendra Place, New Delhi 110008

Original edition © China Science and Technology Press Co., Ltd.
This English edition is authorized by China Science and Technology Press Co., Ltd.

ISBN: 978-1-4878-1160-0

To find out more about our publications, please visit www.royalcollins.com.

# Writing Group Members

## Science Consultant

Zhang Lüqian, Liu Jiaqi

## Writing Committee

Wang Yusheng, Cui Jianping, Qin Deji, Lü Jianhua, Yan Shi, Li Anping, Yang Liwei, Fu Wancheng, Tian Rusen, Li Liya, Li Bowen

## Reviewing Experts

Lu Hong, Li Jin, Zhang Xinmin, Hu Weijia, Xie Yigang

## Editorial Committee

Xu Hui, Zhao Hui, Zhao Jia, Xia Fengjin, Guo Qiuxia

# Contents

# Preface

The Chinese nation is a nation courageous enough to explore and innovate. Throughout the long history, we can see the glorious scientific and technological achievements that countless ancestors have made. These great scientific and technological achievements have not only made a significant contribution to the history of human civilization, but also guided and fueled the emergence and spread of modern world civilization.

Since the emergence of modern science, the Western powers armed with it bullied modern China on the back of their strong ships and artillery. The Chinese people, filled with indignation, have suffered greatly in their fight against it. It is this cruel period of time that has ignited patriotism and the fighting spirit within countless Chinese, that has inspired the Chinese nation to pursue science and democracy, to adhere to self-improvement and self-reliance, and to defend the great motherland with their life.

After the founding of the People's Republic of China (PRC), the Communist Party of China (CPC), as per the needs of the development of China's socialist construction, has timely adjusted strategies, guiding principles, policies, and key tasks of scientific and technological development, and has promoted the coordinated development of science and technology and the economic society. China's science and technical workers continue to grow in strength. Its scientific research power has evolved from weak to strong, gradually forming a scientific and technological system with distinctive Chinese characteristics that cater for the needs of the country's modernization drive.

Seventy years of unremitting struggle leads to 70 years of great changes. In these glorious years, generations of science and technical workers have made great efforts to reach the world's leading level in some important disciplinary fields. They have been repeatedly making great scientific and technological achievements, attracting global attention and making the Chinese people proud. The rapid development of defense-related science and technology has cast a strong backing for strengthening the comprehensive national strength; the rich fruits of civil science and technology have laid a solid social foundation for promoting economic development; breakthroughs and innovations in basic research have enabled China to stride towards becoming a scientific and technological power; the constantly improving basic conditions of scientific research have provided significant support for China to catch up with and surpass the advanced international standards in science and technology; the continuously expanding team of science and technical workers has injected a strong impetus for the development of China's scientific and technological undertakings.

Scientific and technological advancement is an important aspect of economic and social development. And the level of science and technology and innovation capacity are the embodiment of a country's comprehensive national strength. The brilliant scientific and technological achievements made in the past 70 years since the founding of the PRC are the fruits of the hard work and innovation of the Chinese people under the leadership of the CPC. The vast number of science and technical workers who have made these achievements are pioneers of advanced productivity and disseminators of advanced culture. We grow up in the company of the stories of contemporary scientists such as Qian Xuesen, Hua Luogeng, Li Siguang, Wu Wenjun, Wang Xuan, Yuan Longping, and Tu Youyou. In this rapidly changing era, there are heaps of moving stories of science and technical workers. More and more of them have devoted themselves to the development of the country's science and technology without being seen. They have drawn beautiful pictures with their wisdom and knowledge, and written chapters of their magnificent life of patriotic struggle with hard work and dedication. Remain true to our original aspiration and keep our mission firmly in mind. Whether it is the loyal service of overseas Chinese students to their motherland in the 1950s or the current upsurge of returning science and technical talents, a strong China has ignited the aspiration within countless science and technical workers. The rejuvenation of the motherland has become the ideal that science and technology strive for.

The report of the 19th National Congress of the CPC pointed out that from 2020 to 2035, China aims to basically realize socialist modernization on the basis of finishing building a moderately prosperous society in all respects. By then, China's economic strength and scientific and technological strength will have leapt to the forefront of innovative countries. Under the new circumstances in the new times, we firmly believe, under the correct guidance of the Party Central Committee with General Secretary Xi Jinping as the core, we will certainly be able to blaze a path of independent innovation with Chinese characteristics, and the Chinese people who are constantly striving for self-improvement will stand tall in the world with their brilliant achievements in independent innovation.

*Glorious China: The Scientific and Technological Powerhouse Dream* gives a panoramic account of the PRC's development process of science and technology, major achievements in scientific and technological innovation, and science and technical workers who have made outstanding contributions to this end, in order to celebrate the 70th anniversary of the founding of the PRC.

The Writing Group of *Glorious China*
August 1, 2019

# Lay the Foundation

I

On October 1, 1949, the grand opening ceremony was held in Tiananmen Square in Beijing. The People's Republic of China was born! A brand new chapter in the history of China was opened from then on.

On October 1, 1949, the People's Republic of China was founded, marking the end of an old era and the beginning of a new one. Also, China's scientific and technological advancement entered a new period.

When the PRC was newly founded, under the leadership of the CPC, large-scale recovery and construction commenced. In accordance with the national economic construction and development objectives, China's scientific and technological undertakings have also commenced conducting an organized and systematic recovery and construction under the leadership of the CPC and the Chinese government. Starting with basic work, the state approved the establishment of the Chinese Academy of Sciences (CAS), held the first National Congress of Natural Science Workers, adjusted and expanded scientific research institutions, and assembled its own scientific and technological team in the new China.

# A Thousand Things to Do in the New Era

When the PRC was newly founded, under the leadership of the CPC, large-scale recovery and construction commenced. Before the PRC was founded, due to political corruption, economic depression, and frequent wars, the country's science and technology failed to receive due attention and advancement. The existing scientific and technological undertakings were incomplete, understaffed, underfunded, and challenged with adverse conditions, thus lagging far behind the world's advanced level. The CPC was soberly aware that the scientific and technological cause of the PRC must be restored and built in an organized and systematic manner. Therefore, after the founding of the PRC, it immediately placed the development of science and technology under the strong leadership of the CPC and the people's government, beginning to change the decline of the old China to put science and technology on a regular development track. After that, China ushered in a new scientific and technological development era.

## Establishment of the CAS

On October 19, 1949, the Third Meeting of the Central People's Government Commission appointed Guo Moruo as the CAS's first director, Chen Boda, Li Siguang, Tao Menghe, and Zhu Kezhen as deputy directors. On November 1, 1949, in accordance with Article 18 of the *Organic Law of the Central People's Government of the PRC*, the CAS was officially established and directly led by the Government Administration Council. The state initially entrusted CAS with two functions: first, to conduct academic activities with the goal of economic development of the PRC; second, to exercise the administrative tasks of managing all affairs of the natural and social sciences. In 1950, the general policy of national scientific undertakings was put forward:
- Cultivate scientific thinking to eliminate backward and reactionary thinking
- Train sound scientific talents and national construction talents
- Strive to closely coordinate academic research with actual needs, so that science can genuinely serve the state's industry, agriculture, national defense construction, health care, and people's cultural life

On November 1, 1949, the CAS was officially established, with Guo Moruo as its first director.

From June 20 to 26, 1950, the CAS held its first enlarged session in Beijing. Over 100 people attended the meeting. Guo Moruo, Li Siguang, Tao Menghe, and Zhu Kezhen respectively made reports on the policies and tasks, ideological reform, regulations and procedures, and half-year work. After taking over, merging, and adjusting 24 units of the former Central Research Institute, it was decided to establish 13 research institutes, one observatory, and one industrial laboratory.

On June 1, 1995, the founding meeting of the academic divisions of the CAS was held in Beijing.

## The Founding Meeting of the Academic Divisions of the CAS in Beijing

In 1954, the CAS began to prepare for the establishment of the Department of Physics, Mathematics, and Chemistry, Department of Biology and Geography, Department of Technical Science, and Department of Philosophy and Social Sciences, among which 172 scientists from the natural sciences were elected as members of the academic divisions. From June 1 to June 10, 1955, the founding meeting of the academic divisions of the CAS was held in Beijing, announcing the formal establishment of the academic divisions. The meeting summarized the basic experience of scientific undertakings in the past five years and clearly put forward the guiding principles and tasks of scientific projects in the future and the main measures to be taken.

## The Five-Year Scientific Plan of the CAS

When the academic divisions of the CAS were established, it discussed the *Five-Year Plan Outline (Draft) of the CAS*. According to the national objectives and tasks at that time and the policy of the CAS, the Five-Year Scientific Plan of the CAS was proposed, including ten key tasks:

1. Research on the peaceful use of atomic energy
2. Construction and research of new steel base
3. Research on liquid fuel
4. Research on earthquake problems in critical industrial areas
5. Investigation and research on river basin planning and development
6. Investigation and research on tropical plant resources in South China
7. Study and research on natural planning and development in China
8. Research on antibiotics
9. Research on various basic theories in China's transitional state building
10. Analysis of modern China, modern history, and the history of contemporary and modern thought

# Establish a Basic Scientific Research System

With a relatively weak scientific and technological base, after six years of initial construction, the PRC has assembled a scientific research team with the CAS as its main force and has initially set up a scientific and technological system consisting of the CAS, colleges, and universities, industrial departments, and local scientific research institutions. Under the leadership of the CPC and through the hard work of Chinese researchers, various basic disciplines have been improved on their original level, which not only solved some critical problems in the development of the national economy but also laid a solid foundation for a significant number of critical problems in the future development of the national economy. Initially, a relatively complete basic scientific research system was formed. In particular, many scientific seniors who returned to China from abroad brought back information on the world's scientific and technological frontier, laying the foundation for Chinese science.

## Zhang Wenyu Discovered the Subatomic Particle μ

In 1949, during his work at Princeton University, Zhang Wenyu discovered that the slow subatomic particle μ (mu) with negative electric charge produces electrical radiation when interacting with the atomic nucleus. As a result, the subatomic particle μ is named "atom Zhang," and its radiation "radiation Zhang" as well. The so-called subatomic particle μ is a new type of atom formed by the rotation of a negative meson instead of an electron around the nucleus along the stationary orbit, also known as a "strange atom" or "generalized atom." The subatomic particle μ is a basic particle with a mass between an electron and a proton. It is further divided into positive μ, negative μ, and neutral μ according to its electric charge. After Professor Zhang Wenyu discovered the subatomic particle μ, some scientists found that other mesons and hyperons can also form strange atoms. These discoveries, greatly valuable to studying the form, property, and structure of matter, led to the rise of meson physics in the late 1970s.

## Zhou Peiyuan Founded Turbulence Theory Research

Zhou Peiyuan's academic achievements are mainly in two critical aspects of the basic theory of physics: the theory of gravity in Einstein's general theory of relativity and the theory of turbulence in fluid mechanics. In terms of general relativity, Zhou Peiyuan was committed to solving the deterministic solution of the gravitational field equation and applying it to the study of cosmology. In terms of turbulence theory, in the early 1930s, he realized that the turbulent flow field was closely related to the boundary conditions. Later, referring to the method of taking mass as an integral constant in general relativity, he figured out the differential equations satisfied by Reynolds stress, etc. In the 1950s, he used a relatively simple axisymmetric vortex model as the physical image of the turbulence element to explain the homogeneous and isotropic turbulence movement, based on the research on the homogeneous and isotropic turbulence movement. He obtained the binary velocity correlation function and the ternary velocity correlation function in the late and early stages of turbulence decay. Next, he further applied the concept of "quasi similarity" to unify the similarity conditions at the early and late stages of the decay into a physical condition for determining the solution, which was verified by experiments. Thus, the theoretical results of the turbulent energy decay law and Taylor turbulent micro-scale diffusion law from the initial to the later stages of the decay were determined experimentally for the first time in the world.

## Wu Zhonghua Pioneered the Three-Dimensional Flow Theory

In 1944, Wu Zhonghua went to the United States to study. He majored in Engineering Thermophysics at the Massachusetts Institute of Technology, mainly in aircraft research. After graduation, he joined the Lewis Flight Propulsion Laboratory of the National Advisory Committee for Aeronautics. After over three years of research, in 1950 Wu read his paper "A General Theory of Three-Dimensional Flow in Subsonic and Supersonic Turbomachines of Axial-, Radial-, and Mixed-Flow Types on the platform of the American Society of Mechanical Engineers." This theory divides the three-dimensional space inside the impeller infinitely. A complete and real impeller fluid flow mathematical model is established by analyzing each working point in the impeller channel. The blade shape designed per the three-dimensional flow theory has an irregularly curved surface. Such an impeller blade structure can adapt to the real state of the fluid and control the velocity distribution of all fluid particles inside. Therefore, the impeller designed using the three-dimensional flow theory can be installed in the water pump, significantly improving its operating efficiency. The Boeing 747 and the Tristar aircraft from the United States were the largest wide-body civil aviation aircraft at that time. And they were manufactured according to Wu's three-dimensional flow theory. On August 1, 1954, Wu returned to China, taking the post of Deputy Director of the Department of Power Machinery at Tsinghua University. Under his leadership, the first college major in gas turbine engines was founded in the PRC, and it has nurtured many outstanding scientific talents for the country.

## Huang Kun Founded the Polariton Theory

Huang Kun was a pioneer of solid-state physics and the founder of semiconductor technology in China. He published the Huang Kun equation in 1951. It was the first time that photons and transverse optical phonons were coupled to form a new elementary excitation—polariton, which was later confirmed in experiments, making him the first person to propose the internationally recognized concept of phonon polariton. His name is associated with the multi-phonon transition theory, the X-ray diffuse scattering theory, the lattice vibration long wave phenomenological equation, and the semiconductor superlattice optical phonon model. He was committed to the scientific research and education of condensed matter physics. Famous for his diligence, rigor, self-discipline, and tireless teaching, he has trained a significant number of physicists and semiconductor experts for China.

Qian Xuesen (1911–2009), a physicist, world-famous rocket expert, Member of the Academic Divisions of the CAS (Academician), and an Academician of the Chinese Academy of Engineering.

## Qian Xuesen Pioneered Engineering Cybernetics

After World War II ended, Qian Xuesen conducted in-depth observation and research on the rapidly developing control and guidance engineering technologies and made some progress. He introduced the cybernetics idea of Norbert Wiener into his familiar navigation and guidance systems in the aerospace field, thus forming a new discipline: engineering cybernetics. In 1954, his book *Engineering Cybernetics* was published in the United States. From the perspective of technological science, he refined the technical summary of various engineering technology systems into a general theory. The publication of *Engineering Cybernetics* soon attracted the attention of the American scientific community and even the world scientific community, which believes that *Engineering Cybernetics* is a groundbreaking work in this field and another brilliant feat after Norbert Wiener's cybernetics. The book has won an international reputation and has been translated into Russian, German, Chinese, and other languages. As a new technical science, engineering cybernetics provides the basis for the guidance theory of missiles and spacecraft. Qian turned the development experience of China's missile weapons and spacecraft systems into systems engineering theory and applied it to military operations research and socio-economic issues, successfully promoting the development of combat simulation technology and socio-economic systems engineering in China.

# The Rise of Cutting-Edge Technology

After the founding of the PRC, while extensive basic research was being conducted, great efforts were concentrated on developing emerging cutting-edge technologies. In just a few years, China's emerging cutting-edge technologies have advanced from scratch and made considerable progress. Although these new technologies were preliminary and basic, the steps of progress were firm and robust. It marks that the cause of science and technology of the PRC has embarked on a healthy path of development and indicates that China is about to enter a new era of scientific and technological advancement.

## Establishment of Luoxue Mountain Cosmic Rays Research Center

From October to November 1952, Wang Ganchang and Xiao Jian jointly directed the preparation of the first alpine cosmic ray research center in China. They installed the multi-plate cloud chamber brought back by Zhao Zhongyao from the United States and the self-designed magnetic cloud chamber. The research center was finished in 1954 and applied to observe the interaction between cosmic rays and matter.

The earliest cosmic ray field lab in China—Luoxue Mountain Cosmic Rays Research Center, built in Dongchuan, Yunnan Province, in 1954, has made heaps of achievements in researching strange particles and high-energy nuclear interaction, thus laying a foundation for the Chinese nuclear industry.

Hua Luogeng (1910–1985), a mathematician, applied and computational mathematician. Member of the Academic Divisions of the CAS (academician).

## Hua Luogeng Chaired Computer Research

When Von Neumann pioneered and started to design the Electronic Discrete Variable Automatic Computer (EDVAC), Hua Luogeng, who worked at Princeton University, visited his laboratory and used to discuss relevant academic problems with him. Hua returned to China in 1950. In 1952, when college departments throughout China underwent adjustment, he selected Min Naida, Xia Peisu, and Wang Chuanying from the Department of Electrical Engineering of Tsinghua University to establish China's first electronic computer research team in the Institute of Mathematics of the CAS that he was directing at the time. Hua served as the director of the preparatory committee when preparing to establish the Institute of Computing Technology of the CAS in 1956.

## Successful Trial Flight of the Nanchang CJ-5

The Nanchang CJ-5 trainer aircraft was the first primary trainer manufactured by China, and its prototype is the Soviet Yakovlev Yak-18 trainer. Its fuselage had a skeleton welded with alloy steel pipes, a frame-type fuselage skeleton. The front of the fuselage and the engine fairing had aluminum alloy skin, while the rear half of the fuselage had fabric skin. The wings consisted of trapezoidal outer wings and rectangular middle wings. The middle wings had an all-metal structure with two longerons, eight ribs, etc., and were equipped with two 75-liter fuel tanks. The middle wings were connected to the fuselage frame. The outer and tail wings' leading edge, longeron, and rib were made of aluminum alloy and had fabric skin. The aircraft adopted the reliable and easy-to-operate M-11FP5 cylinder air-cooled piston engine. As for the rear three-point landing gear, the main wheel was semi-buried into the middle wings, while the tail wheel was fixed and not retractable. The tandem cabin had a decent view. The aircraft was also equipped with a radio receiver and intercom equipment. On July 11, 1954, the trial flight of China's first Nanchang CJ-5 succeeded.

The trial flight of China's first Nanchang CJ-5 was successful on July 11, 1954.

## Successful Development of Several Nuclear Electronic Instruments

In electronics, the nuclear electronics group headed by Chen Fangyun and Xin Xianjie developed a variety of nuclear electronic instruments. Among them, detectors and scalers began to be mass-produced in 1955, meeting the needs of the uranium geological exploration team, scientific research of modern physics institutes, and university education. Under the guidance of Mei Zhenyue and Zheng Linsheng, the design and development of various β spectrometers and isotope separators also made progress.

## The First Modern Valve Factory in China—the Beijing Valve Factory

On October 15, 1956, the first modern valve factory of the PRC, the state-owned Beijing Valve Factory, was built and put into operation in Jiuxianqiao, an eastern suburb of Beijing. After a year or so, it produced a complete set of thumb-type radio valves, which ended the situation that all radio valves in China had to be imported. The Beijing Valve Factory, having made significant contributions to the development and construction of China's national economy, is known as the "cradle of the Chinese electronics industry."

## Major Breakthroughs in Industrial Manufacturing

During the First Five-Year Plan period, the advanced industrial manufacturing technology of the Soviet Union was adopted to manufacture urgently needed industrial products in China and to set up the technological foundation of Chinese heavy industry. The introduction of 156 complete sets of engineering equipment gradually enabled China to build from scratch industrial sectors such as heavy machine tools, electrical appliances, automobiles, aircraft, ships, and rolling stock and to form an industrial system with relatively complete categories. Researchers have been working diligently to develop new technologies and modern industries in the PRC.

On August 1, 1952, the Qingdao Sifang Locomotive Factory produced the first domestic locomotive of the PRC. This liberation steam locomotive declared to the world that the days when the Chinese were unable to build their own locomotives were over.

On September 14, 1951, the Tianjin Automobile Manufacturing and Distribution Plant completed the trial production of the first Jeep of the PRC.

On October 27, 1953, in Anshan, a steel city, a burning-red billet rolled out of the 1200-degree Celsius high-temperature heating furnace and through the perforator. When the fiery steel pipe head slowly emerged, everybody present shouted and jumped for joy. The first seamless steel pipe of the PRC was made.

Wang Chonglun, a technical innovationist in Anshan Iron and Steel Group Corporation, made great efforts to create and improve new tools and invented the "universal tool matrix."

Ma Xueli, the "king of cutters" of Wuhan Heavy Machine Tool Factory, was working attentively.

# Improve Infrastructure

### Vigorously Harness the Huaihe River

From July to August 1950, the Huaihe River basin suffered catastrophic floods. On October 14, 1950, the Government Administration Council of the Central People's Government of China issued the *Decision on Harnessing the Huaihe River*. The Ministry of Water Resources set up the Huaihe River Harnessing Committee. It formulated the policy of storing flood water and discharging a flood according to different conditions in the upper, middle, and lower reaches. Henceforth, the PRC's first major project of water conservancy construction has begun. By the winter of 1957, 175 large and small rivers had been harnessed, nine reservoirs built, and more than 4,600 kilometers of dikes built, thus significantly improving flood control and discharge capacity.

### Large Yellow River Water Diversion Project—People's Victory Canal

People's Victory Canal was the first large water diversion project built in the Yellow River's lower reaches after the PRC's founding. Its head sluice is located in Jiayingguan Township, Wushe County, Henan Province, and the main canal extends 52.7 km long. The project was started in March 1951, its first phase was successfully completed in March 1952, and its sluice was opened to discharge flood in April of the same year.

The main canal of the People's Victory Canal flows northward along the Beijing–Guangzhou Railway to Xinxiang City and into the Weihe River. It is responsible for irrigation, flood discharge, and Weihe River relief.

The successful opening to traffic of the Sichuan–Xizang and Qinghai–Xizang highways is a remarkable feat in the history of human highway construction.

## Sichuan–Xizang and Qinghai–Xizang Highways Completed and Opened to Traffic

On December 25, 1954, Sichuan–Xizang and Qinghai–Xizang highways were officially opened to traffic simultaneously, ending the history of Xizang having no formal highway. The Sichuan–Xizang Highway starts from Chengdu in Sichuan Province, in the east, crosses the Nujiang River, and climbs the Hengduan Mountains, with a total length of over 2,400 kilometers; the Qinghai–Xizang Highway starts from Xining in Qinghai Province in the north, crosses the Tongtian River and climbs the Kunlun Mountains, with a total length of 2,100 kilometers. The average altitude of the two highways exceeds 4,000 meters, and they converge in Lhasa, the capital of Xizang. The construction of the two highways took five years. Over 3,000 road builders dedicated their lives to their construction, and over 10,000 constructors won awards for meritorious deeds. The successful opening to traffic of the Sichuan–Xizang and Qinghai–Xizang highways is an incredible feat in the history of human highway construction. It has promoted the historic leap of Xizang's social system and the unprecedented development of Xizang's economy and society.

## Vigorously Build Railways

On January 2, 1950, the Central Committee of the CPC approved the construction of the Chengdu–Chongqing Railway and the construction of the Great Southwest. Overcoming all difficulties, one hundred fifty thousand road builders completed it and opened it to traffic on July 1, 1952. It traverses the Sichuan Basin. Along the railway line, there were various products along, which effectively promoted the circulation of materials in the southwest region and played an essential role in production development and economic prosperity.

In 1952, as the Government Administration Council approved, the Ministry of Railways began to conduct field exploration and design for the Baotou–Lanzhou Railway. After two years or so of survey, investigation, demonstration, and analysis, the construction of the Baotou–Lanzhou Railway was started in October 1954, and it was opened to traffic on August 1, 1958.

It traverses three provinces (regions) of Inner Mongolia, Ningxia, and Gansu, and crosses the Yellow River three times, as well as several deserts, with a total length of 990 kilometers.

The Lanzhou–Xinjiang Railway Yellow River Bridge is located in Xigu District, Lanzhou City, Gansu Province. It was the first large railway bridge built on the Yellow River after the founding of the PRC. The construction commenced in April 1954. The bridge was opened to traffic on July 1, 1955, with a total length of 278.4 meters. It met modern bridge standards, and there was no speed limit on the bridge.

Chengdu-Chongqing Railway is the first railroad trunk line in Southwest China and the first railroad built after the founding of the PRC. In 1987, the electrification project of the Chengdu-Chongqing Railway passed the acceptance.

The Baotou–Lanzhou Railway was the first railway crossing the vast Tengger Desert. The illustration shows an array of tractors transported via the Baotou–Lanzhou Railway after it was put to use.

# March towards Science

II

On October 16, 1964, China's first atomic bomb exploded successfully.

When the PRC was just founded, economically, it was still poor and backward. It was an inevitable choice to quickly improve its international status with science and technology. From January 14–20, 1956, a conference on intellectuals was held in Beijing. At the meeting, Chairman Mao Zedong delivered a keynote speech, calling on the entire Communist Party of China (CPC) to study scientific knowledge, unite with non-party intellectuals, and strive to catch up to the world's advanced scientific levels quickly. *The Report on Intellectuals*, read by Premier Zhou Enlai, pointed out, "Science is a determinant factor in all aspects of national defense, economy, and culture." To perform socialist construction, "We must rely on close cooperation between physical and mental labor, and on the fraternal alliance of workers, farmers, and intellectuals." On January 25 of the same year, Chairman Mao Zedong pointed out in his speech at the Supreme State Conference: "The Chinese people ought to have a broad plan to try to change China's backward situation in economy, science, and culture within decades, and quickly reach an advanced international level." At this meeting, he also proposed the policy "may a hundred flowers bloom and a hundred schools of thought contend" for the cause of science, literature, and art.

On January 30, 1956, Premier Zhou Enlai put forward the call of "marching towards modern science and technology" in the political report of the Second Session of the Second Chinese People's Political Consultative Conference. In April of that year, to welcome the world's technical revolution, the Central Committee of the CPC and the State Council co-formulated the *Long-Term Plan for Science and Technology Development* from *1956 to 1967* (referred to as the *Twelve-Year Science and Technology Plan*). *Under the guidance* of the *Twelve-Year Science and Technology Plan* and through the tenacious efforts of Chinese researchers, science and technology in China have undergone fundamental changes and achieved vigorous development.

# Focused Development and Catching up
# Implementation of the *Twelve-Year Science and Technology Plan*

When the PRC was newly founded, a thousand things waited to be done. The country's economic construction had just resumed and was in a poor situation. China must constantly strive for self-improvement to rapidly develop the national economy and adapt to the situation of international struggle. In 1956, under the leadership of Premier Zhou Enlai, the State Council established the Planning Commission, mobilized hundreds of scientists from all walks of life and disciplines to participate in the preparation of the plan, and invited 16 Soviet scientists from all disciplines to China to help understand the international scientific and technological standards and development trends. After seven months of repeated revisions, in December 1956, with the approval of the Central Committee of the CPC and the State Council, the *Twelve-Year Science and Technology Plan* was promulgated. This was China's strategic plan for science and technology, and it opened the prelude to the country's march toward science.

The *Twelve-Year Science and Technology Plan* aimed to introduce the world's most advanced scientific achievements to China's science and technology departments, national defense departments, production departments, and education departments as quickly as possible and to make up the categories that the Chinese scientific community most urgently needed for national defense construction as soon as possible. This has found a specific form of organization and realization for science to serve national construction, significantly improved the efficiency of scientific research, accelerated the pace for China to catch up with the advanced international level of science and technology, and contributed to the achievement of "missiles, nuclear bombs, and the artificial satellite" in the following ten years or so. Ultimately, it led to the establishment and development of new technology fields such as computers, automation, electronics, semiconductors, new materials, precision instruments, etc.

## (1)  Achievements in National Industrialization and National Defense

### Modernization

In the early 1960s, China encountered a severe predicament. The Soviet Union withdrew all its experts. Chinese scientists and technicians were independent and self-reliant. They kept tackling critical scientific research problems and continued developing missiles and atomic bombs. With the support of Mao Zedong and Zhou Enlai, among other leaders, a series of measures were taken to highlight the key points, prioritize tasks, organize nationwide cooperation, vigorously tackle technical difficulties such as new raw materials, electronic components, instruments and meters, precision machinery and large equipment, and further adjust intellectual policies. It took only five years to develop a variety of missiles and atomic bombs successfully. Soon after, hydrogen bombs laid the foundation for developing long-range rockets, artificial satellites, and nuclear submarines successfully. At the same time, remarkable achievements have been made in developing conventional weapons and equipment and civil scientific research projects.

In 1958, China built its first atomic reactor.

## Atomic Energy Development

Atomic energy, also known as nuclear energy, is released when the atomic nucleus changes. There are two kinds of nuclear changes: heavy nuclear fission and light nuclear fusion, both of which can release tremendous energy. In addition to atomic energy for military purposes, every country in the world was developing the application of nuclear power in economic fields. The Chinese nuclear energy industry was founded in 1955. With less investment, China has figured out a development path suited to its national conditions and has made remarkable achievements in a relatively short period.

In May 1956, China began to build its first atomic reactor, officially put into operation two years later. The primary purpose of this reactor was to make isotopes and conduct scientific research. It used uranium as fuel and heavy water as moderator and heat conductor, hence the name of an experimental heavy water reactor. The completion of the reactor marked that China had entered the era of atomic energy.

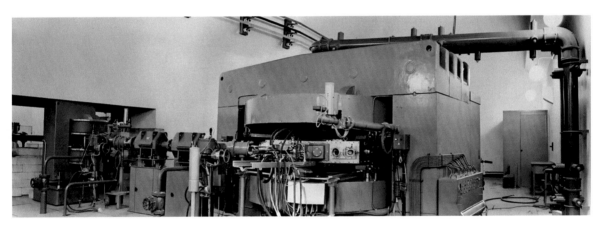

China's first cyclotron.

## The First Cyclotron in China Was Built

On June 30, 1958, China's first cyclotron was completed. A cyclotron is a kind of resonant accelerator in which particles move along an arc orbit. In a constant strong magnetic field, particles are accelerated many times by a high-frequency electric field with a fixed frequency to obtain sufficient energy. Accelerators can be used in nuclear experiments, radioactive medicine, radiochemistry, radioisotope manufacturing, non-destructive flaw detection, etc.

## Research and Development (R&D) of Rockets and Missiles

The *Twelve-Year Science and Technology Plan* made it clear that Chinese rocket technology must walk the path of independent development within 12 years. In September 1958, China's first high-altitude sounding rocket, 6 meters high, spat a long string of red flames and soared into the sky in the wilderness of Baicheng in Jilin Province, lifting the historical curtain of China's time in space. The rocket was named Beijing No. 2 and was successfully developed and launched by teachers and students at the Beijing Institute of Aeronautics and Astronautics (now Beijing University of Aeronautics and Astronautics). On February 16, 1960, China's first liquid rocket successfully took off.

Missile is short for a guided missile. It is a weapon that is propelled by its power device, directed and controlled by the guidance system, and led straight to its target. The development of missiles was the need for China's national defense modernization. Qian Xuesen, a physicist, made outstanding contributions to the creation and development of China's rocket, missile, and aerospace undertakings. He was the initiator of the country's system engineering theory and application research, thus known as the "father of missiles" in China.

Qian Xuesen returned from the United States to his motherland in 1955.

Qian Xuesen instructed the work at the missile launch site.

China's first strategic missile was successfully launched.

## Large General-Purpose Computer—The Successful Development of Computer 119

The electronic computer is an epoch-making invention. The development of electronic computers in the United States was earlier, while this technology was blocked in China. The *Twelve-Year Science and Technology Plan* listed "the establishment of computing technology" as one of the four national emergency measures. In terms of organization, the policy of "centralization before decentralization" was adopted; in terms of scientific research, the policy of "imitation first, innovation after, imitation for innovation" was determined; the guiding principle of organizing training courses, continuing education, cooperating and sending talents abroad to study was applied to speed up the training of cadres. In 1964, under the leadership of Wu Jikang, the Institute of Computing of the CAS successfully developed the large general-purpose computer 119. It adopted a high-speed logic circuit composed of valves and crystal diodes and was equipped with a 16,000-character magnetic core memory. Some external devices could work in parallel with the central processor to provide users with BCY algorithm language. The Computer 119 had a complete instruction system, high computing speed, large storage capacity, strong problem-solving ability, and convenient, stable, and reliable operation. It has completed a great many atomic energy, weather forecast, and other computing tasks. The successful development of Computer 119 marks a new stage in China's self-relied computer development. It is precisely because we have developed our electronic computers and nuclear and aerospace technology that our national defense cutting-edge science can stand strong and capable in the world.

Computer 119, the first large general-purpose digital electronic computer independently researched, designed, and manufactured in China, won the first prize in the National Industrial New Product Exhibition in 1964.

## R&D of Semiconductors

During this period, due to the blockade of Western countries on China, it failed to grasp the international research trends on semiconductors in time. In addition, the dearth of advanced instruments and equipment made scientific research face enormous difficulties. For example, the development of silicon high counter-voltage transistors. In the late 1950s, to overcome difficulties, a group of young talents endured countless failures in the laboratory before achieving success at last. Some of their semiconductor research achievements have approached the advanced international level. It mattered a great deal for the development of radio electronics and automation technology to be able to manufacture diodes and triodes that are small, long-lived, stable, and reliable. Unfortunately, Chinese scientists did not foresee the development of integrated circuits and large-scale integrated blocks at that time, so China's research in this area fell behind the international standard for a decade.

In 1962, Lin Lanying developed the first gallium arsenide single crystal in China, reaching the highest international level.

## R&D of Radio Electronics

Wireless communication technology is not only a key technology in national defense construction but also an essential technology in economic construction. The importance of radio electronics lies not only in communication but also in civil technology and modern national defense technology. Industry, agriculture, medicine, and health are inseparable from radio electronics. Radio electronics are also necessary for national defense technology, such as radar and automatic artillery design and command. The development of radio electronics is a task of great significance. During the implementation of the *Twelve-Year Science and Technology Plan*, the Institute of Electronics of the CAS has formed four leading fields: microwave imaging radar and its application technology, microwave devices and technology, high-power gas laser technology, and microsensor technology and systems. Scientists of the older generation have made outstanding contributions to China's radio industry.

In November 1956, under the leadership of semiconductor experts Wang Shouwu, Lin Lanying, and Wu Erzhen, China's first germanium alloy transistor was made in Beijing North China Radio Components Research Institute. As a pioneer of radio electronics in China, Chen Fangyun participated in constructing China's satellite measurement and control system in 1964 and contributed to launching Chinese artificial satellites. The microwave unified measurement and control system he proposed and helped complete has become the leading equipment to support the launch of Chinese communications satellites. This project won the special prize of the National Science and Technology Progress Award in 1985.

As one of the founders of radio electronics in China, Meng Zhaoying has taught at the university for over 60 years, making outstanding achievements in talent training, laboratory construction, and textbook compilation. He has significantly contributed to scientific research in microwave electronics, spectroscopy, and cathode electronics, among other fields.

The illustration shows the universal milling machine with a rock arm produced by Nantong Machine Tool Factory. It was built as the industry of the PRC rose. It was where China's first universal milling machine with a rock arm, the first CNC milling machine, the first small vertical machining center, and the grassland machinery industrial base were born.

## Developed the First-Generation CNC Machine Tool

Automation technology is one of the science technologies that developed rapidly and greatly impacted the 20th century. Modern automation technology is a new type of productivity and one of the principal means to directly create social wealth. It plays a great role in promoting human production activities and material civilization. Therefore, automation technology has received extensive attention from countries worldwide and been increasingly applied. The Central Committee of the CPC decided to develop this technology because it foresaw that the future development of industry would inevitably be automation. This saves not only extensive labor but also is necessary to ensure high-quality products. In future warfare, there must be automatic offensive and defensive equipment; otherwise, the country cannot adapt to modern warfare with high sensitivity and quick response in the future.

In the 1960s, China developed the first generation of CNC machine tools. The development and production of CNC systems in China have made significant progress, basically mastering key technologies, establishing CNC development and production bases, cultivating a group of CNC talents, initially forming our own CNC industry, and driving the development of electromechanical control and transmission control technologies.

## Developed the First Large-Scale Electron Microscope

Shanghai Precision Medical Equipment Factory and Changchun Institute of Optics and Precision Machinery of the CAS have cooperated since 1960 to co-research and co-develop an electron microscope, whose trial production succeeded on August 2, 1965. It was the first first-grade large-scale electron microscope in China. It had a magnification of 200,000 times and a resolution of 7 angstroms and could be widely used in scientific research and industrial and agricultural production. It was designed and manufactured by China itself using only domestic raw materials. Its birth marks that China's microscope manufacturing technology and corresponding optical research level have entered the advanced international ranks.

The 200,000 times high-resolution electron microscope made by several industrial and scientific research institutions in Shanghai contains tens of thousands of components. It requires high-end manufacturing technology and many unique materials.

## China's First Ruby Laser

In September 1961, China's first ruby laser was made at the Changchun Institute of Optics and Precision Machinery of the CAS, only one year later than in foreign countries, but it had a unique structure. Its birth shows that Chinese laser technology has entered the advanced international ranks, opening up a new field for Chinese laser technology. Henceforth, laser technology has been increasingly applied in national defense and industrial and agricultural production.

China's first ruby laser.

# (2)  The Survey of Natural Conditions and Resources in China

China has a vast territory, with a total land area of about 9.6 million square kilometers, including tropical, subtropical, temperate, and cold zones; there are the highest mountains and plateaus in the world, as well as vast and fertile plains; the water area of the inland and border seas covers about 4.7 million square kilometers, and the tortuous coastline stretches a total length of over 18,000 kilometers. China has excellent natural conditions and abundant natural resources. To make full use of and develop them promptly, we must conduct a series of investigations and research to grasp the changing laws of natural conditions and the distribution of natural resources and thus determine the direction of their utilization and development; on this basis, we will study the prospects for the development of the national economy of each region and the entire country, as well as the plans for the rational allocation of industry and agriculture.

Zhu Kezhen is the initiator, founder, and leader of China's comprehensive investigation cause. Per the *Twelve-Year Science and Technology Plan*, he summarized the comprehensive national investigation into five tasks:

1.  Research on the comprehensive investigation and development plan of the Xizang Plateau and the Hengduan Mountains in Kangdian
2.  Research on the comprehensive investigation and development plan in Xinjiang, Qinghai, Gansu, and Inner Mongolia
3.  R&D of unique biological resources in tropical areas
4.  Thorough investigation of water resources of vital rivers
5.  Natural and economic regionalization in the country

### Achievements in Natural Resources Investigation

There is a good tradition of scientific investigation in China. According to the task of investigating the natural conditions and natural resources of remote areas in China stipulated in the *Twelve-*

The action center of Hengduan Mountain glaciers is in Gongga Mountain, China's most concentrated and most significant area of modern marine glaciers and the best preserved and developed area of Quaternary glacial relics in the east of Qinghai-Xizang Plateau. The illustration shows glacier researchers riding toward the west slope of Gongga Mountain.

*Year Science and Technology Plan*, the CAS has set up a comprehensive investigation committee to lead and organize various comprehensive investigations. 11 comprehensive investigation teams have been assembled to conduct comprehensive investigations in Inner Mongolia, Northeast, Northwest, Southwest, Southeast, and South China, Thousands of researchers were conducting the inquiry simultaneously every year. The comprehensive examination of natural resources in that period has made palpable achievements. They covered an area of 5.7 million square kilometers. Extensive investigations and studies have been made on many major research topics related to the transformation of nature, including water and soil conservation, desert control, grassland improvement, etc., thus providing a large amount of scientific data and a basis for the national development and utilization of natural resources, the formulation of national economic development plans and regional development plans; and they have promoted the development of resource research and investigation.

The Qaidam Basin is a unique natural treasure in China. With over 200,000 square kilometers, the basin abounds in rich mineral resources. The Dameigou Section in Qaidam Basin is a typical section of Jurassic in China, with a total thickness that exceeds 1,100 meters.

Under severe environment,Chinese scientists have made contributions to the scientific investigation of Mount Everest.

## Remnants of Glaciation Investigation

Mineral resources are an essential condition for one country's economic construction. China has a vast territory and abundant mineral resources. Most of the known minerals in the world are available in China, among which the reserves of tungsten, rare earth, coal, iron ore, etc., rank among the world's top. Li Siguang, a geologist, founded China's Quaternary Glacier Theory and Geomechanics Theory. His discovery and theoretical practice have brought epoch-making significance to the exploration of mineral resources and the development and utilization of engineering geological environmental resources in China.

## (3) Breakthroughs in Major Technical Problems in Industrial Construction

Realizing national socialist industrialization is the objective requirement and necessary condition for a country to become independent, prosperous, and strong. In 1956, the Central Committee of the CPC explicitly proposed the policies of establishing a separate and complete industrial system. These policies have far-reaching significance for China to adhere to its independent position in the wake of drastic changes in international relations. During the implementation of the *Twelve-Year Science and Technology Plan* in 1956, First Automotive Works (FAW), the first factory to produce motor trucks in China, was completed and put into operation; the first aircraft factory in China successfully trial produced the first jet aircraft; Shenyang First Machine Tool Factory, the first machine tool factory in China, was completed and put into operation; Beijing Valve Factory, which mass-produced valves, was officially put into operation. In 1957, the Wuhan Yangtze River Bridge that connects the south bank and the north bank of the Yangtze River was completed. The Kangding-Xizang and Xinjiang-Xizang highways were built and opened to traffic one after another.

Many technical problems in industrial construction have been solved. Chinese scientists and technicians have been able to design and develop projects, big and small, including a 1.5-million-ton iron and steel complex, an oil refinery with an annual output of one million tons, a 650,000-kilowatt hydropower plant, and an electrified railway. China's industrial production capacity has increased significantly; many industrial and mining enterprises have been founded on the mainland, which has initially improved the irrational distribution of over-centered industries towards the coast. Compared with the growth rate of other countries during the period of their industrial take-off, China was among the best. Thanks to the arduous struggle of the whole CPC and every Chinese, China's socialist industrialization is steadily marching forward. China's independent construction of a 12,000-ton hydraulic press and the achievements of Daqing oilfield exploration are the highlights of industrial accomplishments in this period.

Jie Fang trucks manufactured by the FAW.

## FAW Completed and Put into Operation

In December 1950, Chairman Mao Zedong visited the Soviet Union. The two sides agreed that the Soviet Union would provide comprehensive assistance to China in building its first motortruck factory. After a year or so of research and comparison of several schemes, it was decided in 1951 to locate the FAW in the suburb of Changchun, Jilin Province. On October 15, 1956, the FAW was officially completed and started mass production. The plant received a total investment of 650 million yuan and had an annual output of 30,000 motor trucks. The Jie Fang trucks manufactured by the FAW were designed and manufactured based on the model of the USSR's Studebaker-US6 with improvements on the part of the structure according to the actual situation in China. This truck had an engine with 90 horsepower and six cylinders. It could run at a maximum speed of 65 kilometers per hour and had a load capacity of four tons. It was suitable for the load capacity of roads and bridges in China then and could be modified into various vehicles suitable for different special purposes. The first batch of these trucks was proved to have good performance and meet the design requirements after the driving test.

The FAW carried out a campaign of technical innovation and technological revolution. The illustration shows the automatic cylinder block processing line. It required only two-person management to complete 34–60 processes, significantly improving work efficiency.

## China's First Jet, the Shenyang J-5, Successfully Flew for the First Time

On July 19, 1956, the first jet interceptor and fighter aircraft in China were manufactured by Shenyang Aircraft Manufacturing Company (formerly known as Factory 112)—Zhong 0101 arrived at Shenyang Yuhong Airport and successfully took off. The trial verified that at the maximum speed and altitude, the performance and data of the special equipment and engines of Zhong 0101 were all up to the requirements of the flight test program. On September 8, the National Acceptance Committee held an acceptance signing ceremony in Factory 112 and named the aircraft Type 56 (later called J-5 according to its series). J-5 aircraft was mainly used in daytime interception and air combat, with specific attack capability. Its improved J-5 A had radar in its nose for night interception and air combat. The first flight of J-5 ended the history that China could not manufacture jet fighter aircraft.

J-5 was derived from the Soviet Mikoyan-Gurevich MiG-17. On July 19, 1956, the J-5 prototype successfully flew for the first time, and it was mass-produced in September of the same year. The production stopped in September 1959, with 767 aircraft produced. The illustration shows the J-5 aircraft for viewing in the park.

## "Anshan Iron and Steel in the South"—Xinyu Iron and Steel Plant

The construction of Jiangxi Xinyu Iron and Steel Plant (from now on referred to as Xinyu Iron and Steel) took place in the specific era of "catching up with Britain and surpassing the United States" to become a powerful iron and steel country. It was proposed to build a steel industry layout of "three big, five middle, and eighteen small" iron and steel plants back then. During the Second Five-Year Plan (1958–1962), the Ministry of Metallurgical Industry decided to build Xinyu Iron and Steel into a steel complex that produces two million tons of iron and 1.5 million tons of steel annually, and called it "Anshan Iron and Steel in the South." In 1960, two 255-cubic-meter blast furnaces were built and operated in Xinyu Iron and Steel. The plant adopted old backward methods to build five small blast furnaces and 14 simple small coke ovens and conducted vigorous pig iron smelting in blast furnaces.

In 1961, when the number of cadres and workers was sharply reduced as some were dismissed, the plant did not give up but chose another way of converting the blast furnace to produce ferromanganese. Through hard work, Xinyu Iron and Steel has overcome the technical difficulties, successfully transforming the blast furnace gas purification system and converting the blast furnace to smelt ferromanganese. Among them, Yuanhe blast furnace ferromanganese, rated as a national high-quality product, won a national silver medal. At the beginning of 1964, Xinyu Iron and Steel was transformed into a special Jiangxi Steel Plant, mainly producing raw military materials. The steel plant operation has brought temporary prosperity to the steel industry and contributed to the aviation industry. Shanfeng steel wire products occupied 60% of the domestic market and were known as "a wonder" in the special steel industry.

The illustration shows an old backward way Xinyu Iron and Steel Plant being built, and the PLA was helping to build a steel base.

## Wuhan Yangtze River Bridge and Nanjing Yangtze River Bridge Were Built Successively

On September 1, 1955, the construction of the Wuhan Yangtze River Bridge commenced, and it was opened to traffic on October 15, 1957. The bridge was initially built with the help of the Soviet government at that time. Later, because the Soviet government withdrew all experts, Mao Yisheng led the unfinished project. Wuhan Yangtze River Bridge was the first bridge with dual uses of railway and highway built by China on the long Yangtze River, connecting the three cities (Wuchang, Hankou, and Hanyang) of Wuhan, as well as the Beijing–Wuhan Railway and the Guangdong–Wuhan Railway separated by the Yangtze River, forming a complete Beijing–Guangzhou Railway. The main bridge is 1,155 meters long. With the bridge approaches at both ends, it stretches 1,670 meters long. The bridge body has continuous triple beams, each of which has three holes. There are eight piers and nine holes in total. The span of each hole is 128 meters, making it easy for huge ships to pass all year round. There are bridgeheads with national style on both ends of the main bridge, each 35 meters high, seven floors from the ground floor hall to the top pavilion, which can be accessed through elevators. The entire bridge is extremely magnificent, and its ancillary buildings and various decorations are well coordinated.

Nanjing Yangtze River Bridge is a double-layer bridge with both highway and railway designed and built by Chinese scientists and technicians independently. It was one of the critical achievements of China's economic construction in the 1960s. The completion of this bridge is a milestone. The bridge consists of a main bridge and bridge approaches. The main bridge has nine piers and ten spans, with a length of 1,576 meters and a maximum span of 160 meters. It has a navigation clearance width of 120 meters and a navigation clearance height of 24 meters above the design maximum for water-level navigation. As a result, a 5,000-ton seagoing ship can easily pass. The bridge was fully completed and opened to traffic in 1968, connecting the original Tianjin-Shanghai and Shanghai-Nanjing railways to the Beijing-Shanghai railway.

Mao Yisheng (1896–1989), a bridge scientist and member (academician) of the Academic Divisions of the CAS, led the construction of the Wuhan Yangtze River Bridge.

In 1957, the Wuhan Yangtze River Bridge was completed.

On December 29, 1968, the Nanjing Yangtze River Bridge was completed and opened to traffic.

## The Yellow River Sanmen Gorge Water Control Project

Sanmen Gorge Water Control Project, known as the "First Dam of the Yellow River," is the first large-scale water control project focused on flood control and comprehensive utilization built on the Yellow River after the founding of the PRC. Its controlled river basin coverage is 688,400 square kilometers, accounting for 91.5% of the total river basin, controlling 89% of the water and 98% of the sediment inflow of the Yellow River. The project was built in 1957 and basically completed in 1960. The main dam is a concrete gravity dam with a height of 106 meters and a length of 713.2 meters. The installed gross capacity of this project is 400,000 kilowatts, bringing into play huge social and economic benefits.

Sanmen Gorge Water Control Project on the Yellow River.

Shi Changxu (1920–2014), member of the Academic Divisions of the CAS (Academician), academician of the CAE (China Academy of Engineering), and expert in metallurgy and materials science (second from the right).

In 1954, the story of Shi Changxu, a Chinese student who was studying in the United States and striving to return home, was published in American newspapers.

## Shi Changxu Led the Research of Cast Superalloy

Shi Changxu was not only a master in the field of materials science and technology in China but also an outstanding manager and science and technology strategist who promoted the development of materials science in the country. In the late 1950s, superalloys were essential materials for the development of aviation, aerospace, and atomic energy industries. Shi proposed the strategic guiding principle of vigorously developing iron base superalloys based on the fact that China was short of nickel and chromium and blocked by capitalist countries. In order to overcome the weakness of poor heat resistance of general iron base superalloys, he and others abandoned the conventional practice of high titanium and low aluminum in iron base superalloys when designing compositions, but accordingly increased the aluminum content, thus developing the first iron base superalloy GH135 (808) in China, which replaced the nickel-base superalloy GH33 at that time as the turbine disk material of aero engines. Shi has made outstanding contributions to building an independent industrial system and national economic system in China.

## First Electrified Railway

Baoji–Chengdu Railway was the first formidable railway built after the founding of the PRC, and the section from Baoji to Fengzhou was full of hardships and dangers. In the section with complex terrain, a steam locomotive is used to draw forward the train, hence its small traction weight, slow speed, and low transportation efficiency. Therefore, when building the Baoji–Chengdu Railway, the

Baoji-Chengdu Railway is the first electrified railway in China. All locomotives running on this railway are independently designed by China itself.

Ministry of Railways decided to adopt electric locomotive traction for the section from Baoji to Fengzhou. In this way, the limit gradient of the route could be increased from 20% to 30%. The route could be shortened by 18 km, the tunnel by 12 km, and the construction period by one year. The electrified railway of the Baoji-Fengzhou section was designed and built by China itself, and thousands of materials and equipment used were produced domestically.

From the very beginning, the Chinese electrified railway has chosen the most advanced current system in the world, avoiding walking the old path of developing countries from direct current to alternating current. After two years of hard work, the first electrified railway in China was completed in May 1961 and started trial operation with a domestic Shaoshan Type 1 electric locomotive. It was officially delivered for operation on August 15, 1961. Henceforth, Sichuan, where transportation had been a predicament since ancient times, has become an easy destination. The completion of the 91 km Baoji–Fengzhou Electrified Railway has promoted the development of Chinese electrified railways.

## Domestically-Built 12,000-Ton Hydraulic Press

In May 1958, during the Eighth National Congress of the CPC, Shen Hong, then Vice Minister of the Ministry of Coal Industry, wrote a letter to Chairman Mao Zedong, proposing to use Shanghai's original machine manufacturing capacity to independently design and manufacture a 10,000-ton-level hydraulic press, so as to change the situation in which large forgings had to be imported. This proposal was supported by Chairman Mao Zedong. Therefore, the relevant central departments assigned the task of manufacturing 10,000-ton-level hydraulic presses to Shanghai Jiangnan Shipyard. In 1959, the shipyard set up a 10,000-ton-level hydraulic press team. Under the leadership of chief designer Shen Hong, technicians and workers overcame a series of difficulties by dividing a giant difficulty into dozens of smaller difficulties. After four years of endeavor, the 10,000-ton-level hydraulic press was built and put into production on

Shen Hong (1906–1998), an expert in mechanical engineering, was a member of the Academic Divisions of the CAS (academician).

The 12,000-ton hydraulic press manufactured under the leadership of Shen Hong in 1962.

June 22, 1962. It was 23.65 meters high, 33.6 meters long, and 8.58 meters wide. The whole machine is composed of over 44,700 components. The total weight of the machine body is 2,213 tons, of which the largest component is the lower beam, which weighs 260 tons. The working fluid pressure is 350 Pa, thus able to forge 250-ton steel ingots. It was a key piece of equipment in the heavy machine manufacturing industry.

## Fruits of Daqing Oilfield Exploration

In 1955, China launched an unprecedented petroleum exploration, forming a complete theory of continental oil generation and accumulation. Based on the country's geological structure characteristics and continental oil generation theory, Karamay Oilfield has been found in Junggar Basin, Xinjiang, and oil and gas fields have been found in Jiuquan, Qaidam, Tarim, Sichuan, Ordos, among other basins, fully demonstrating the oil and gas prospects of continental strata.

In 1958, the Ministry of Geology and the Ministry of Petroleum shifted the focus of oil exploration to the eastern region that had been determined by foreign experts as "without crude oil." Regional exploration has been performed in several large basins in Northeast China and North China. On September 6, 1959, industrial oil flow was discovered in the continental sedimentary rocks of Songliao Basin in Northeast China. This was a major achievement in China's petroleum geology undertaking. It happened to be the 10th National Day (*guoqing* in Chinese). Therefore, this oilfield was named Daqing.

Wang Jinxi (1923–1970), a model worker on the oil front.

PLA soldiers in the Daqing Oil Battle.

On February 20, 1960, the Central Committee of the CPC decided to start a campaign on oil exploration and development in Daqing, Heilongjiang Province.

The Daqing Oilfield, located in the west of Heilongjiang Province and the north of the central depression of Songliao Basin, is the largest comprehensive oil production base in China and one of the world's largest oilfields. Since the discovery of the Daqing Oilfield in China, an unprecedented oil battle was immediately started in Daqing. Wang Jinxi led drilling team 1205 from the Yumen Oilfield in the northwest to join the struggle. With no roads and insufficient vehicles, he managed to lead the whole crew to transport the drilling equipment to the construction site, worked relentlessly for five days and five nights, and drilled the first gusher in Daqing.

In the subsequent ten months, Wang led both drilling team 1205 and drilling team 1202, under extremely difficult circumstances, to achieve the miracle of an annual drilling depth of 100,000 meters.

Daqing oil workers, represented by Wang Jinxi, worked diligently and started from scratch. In only two years, the Daqing Oilfield was basically built. At the end of 1963, Premier Zhou Enlai announced in the government work report that "the era when Chinese had to import foreign oil will soon be gone for good."

# (4)  Solving Problems Related to Agricultural Construction

The *Twelve-Year Science and Technology Plan* proposes that in order to rapidly develop agriculture, forestry, animal husbandry, aquaculture, and sericulture, it is necessary to study ways to increase the output per unit area and expand the area (such as reclamation) to give play to the potential of increasing production of labor and land. Meanwhile, in order to realize agricultural mechanization, we must, within 12 years, well select and improve agricultural machinery, forestry machinery, animal husbandry machinery, aquatic machinery, and so on, and formulate a complete set of technical plans for mechanized farming, cultivation, forest harvesting, timber transportation, livestock breeding and management, and fishing, etc. Agriculture, forestry, and animal husbandry are closely linked, and the combination of the three is of great significance to continuously improve their output.

## Completed the National Cultivated Land Survey and Planning

As per the principles formulated in the *Twelve-Year Science and Technology Plan*, through the endeavor of agricultural scientists and technicians, the national survey of cultivated land soil has been preliminarily completed, crop varieties in various regions of the country have been sorted and identified, and a great number of improved strain that could be widely promoted have been found, such as the 169 improved strain of eight crops, including rice, wheat, cotton, corn, etc. We have basically mastered the occurrence rules of 11 crop diseases and pests and put forward many effective control and prevention methods, such as basically eliminating locusts. Meanwhile, we have developed and improved many vaccines against diseases of livestock and draught animals and developed a number of machines and tools suitable for agricultural production conditions in various regions. The research on the improvement of livestock varieties, migration law of fish, investigation of fishery resources, rapid growth and high yield of trees, rubber planting, and so on has harvested fruitful results.

## Ding Ying Laid a Foundation for Rice Cropping in China

In 1926, Ding Ying found wild rice in the eastern suburb of Guangzhou and then proved that China is one of the places of origin of rice cropping; from 1955 to 1959, he performed an in-depth study on the growth and decline of tillers, the development of young ears, and the filling of grains, which are closely related to the formation of rice yield. Through the research, some common results have been found from the relationship between technical measures and ear number, grain number, and grain weight, which provides a theoretical basis for the manual control of seedlings, plants, ears, and grains to achieve the planned yield target; in addition, the rationality of technical measures can also be tested according to what happens in the growth and of rice, which provides a scientific method for summarizing the experience of the masses in cultivating rice, and is beneficial to the development of agricultural production, scientific research, and education.

Ding Ying divided China's cultivated rice into two subspecies, indica, and round-grained non-glutinous rice, and classified the cultivated varieties according to the levels of indica/ round-grained non-glutinous rice, late/early, water/land, and sticky/glutinous from the perspective of ecology; consequently, he has cultivated many excellent varieties for production and contributed to improving yield and quality.

## Qiu Shibang, No.1 in Controlling Locusts in China

China used to suffer from serious locust disasters. The horror of locust disasters is that they are rampant. In Chinese history, many have fought hard against this pest. However, it was not until after the founding of the PRC that this issue was effectively addressed. Honored as "No.1 in Controlling Locusts in China," Qiu Shibang, having long been engaged in agricultural insect research, gradually developed a set of mature methods for locust eradication and control. In the 1940s, "666" powder was first used to control locusts in China, and DDT was used to control pine caterpillars; in the 1950s, China developed the locust detection technology of egg detection, hopper detection, and adult insect detection for the first time, and proposed to use poison bait to control locusts; in the 1960s, the control technology of corn borer was developed and applied nationwide; in the 1970s, it was advocated to implement integrated control of crop pests and focus on biological control research. A set of methods suitable for the breeding of lacewings were developed in rural China, creating conditions for biological pest control. Qiu, having made a great endeavor to prevent and control pests in China as well as great achievements, was awarded the Patriotic Harvest Award by the Ministry of Agriculture in 1954; in 1978, he won the Advanced Individual Award of the National Science Conference; in 1979, he was bestowed the title of National Model Worker.

After the founding of the PRC, the prevention and control of pests have improved greatly. Qiu Shibang (left), an insect pest expert, was a researcher at the Institute of Plant Protection. He has made important achievements in locust research and certain contributions to the eradication of locusts in China. Qiu Shibang (1911–2010) was a modern Chinese agricultural entomologist and academician of the CAS.

## The Birth of Dongfanghong Tractor

In 1955, construction of the First Tractor Factory, one of the 156 key projects during the First Five-Year Plan period of the PRC, commenced in Luoyang. In 1958, the first Dongfanghong high-power crawler tractor of the PRC was made in Luoyang, opening a new page in China's agricultural history. On November 1, 1959, Vice Premier Tan Zhenlin solemnly announced to the world after attending the inauguration ceremony of the First Tractor Factory: "The era in which Chinese no longer need cattle to plow has begun." Thereafter, the prelude of agricultural mechanization in the PRC officially sounded in Luoyang. In the following two decades, Dongfanghong tractors completed over 60% of machine-plowing work in China.

Dongfanghong tractors manufactured in 1958.

## (5) Facilitate Significant Progress in Health Care

The *Twelve-Year Science and Technology Plan* makes the following requirements for health care: actively prevent and treat various major diseases and constantly improve people's health; produce and develop new antibiotics, drugs, biological products, plasma products, and their substitutes; conduct medical research of radioisotope and clinical application; strengthen the collation, research, and development of traditional Chinese medicine, improve environmental health, provide balanced nutrition, promote reasonable physical exercise, and actively improve health, prolong life, and enhance labor capacity.

During the implementation of the *Twelve-Year Science and Technology Plan*, the Chinese healthcare industry has made considerable progress, especially in several aspects of clinical medicine, which has approached or reached the advanced international level.

## Development of Live Inactivated Vaccine for Polio

Poliomyelitis is an ancient disease, often referred to as polio, which poses a serious threat to children's health. It can cause bone growth abnormalities in children. Vaccination with live attenuated poliomyelitis vaccine is an effective way to prevent the disease. Since the mid-1950s, R&D of this vaccine has started. In March 1960, the first batch (5 million doses) of domestic live attenuated polio vaccine was successfully developed and used for childhood immunization. In 2000, the World Health Organization certified that China had achieved the polio-free goal.

## Development of Antibiotics

Before the founding of the PRC, all antibiotic drugs in China were imported. After 1949, antibiotic laboratories were established in Beijing and Shanghai, among other cities. On May 1, 1953, penicillin was officially put into production in Shanghai's No. 3 Pharmaceutical Factory, opening up the Chinese antibiotic

In 1960, China successfully developed a live attenuated polio vaccine. The illustration is Gu Fangzhou (1926–2019), the father of the Chinese polio vaccine, who was known as the "Grandpa of Sugar Pills." He dedicated his life to achieving the eradication of polio in China and maintaining polio-free for good.

industry. On June 3, 1958, North China Pharmaceutical Factory, the largest antibiotic enterprise in China, was completed and put into operation. Subsequently, more than 60 varieties of medicine, including chloromycetin, erythromycin, kanamycin, neomycin, and bacitracin, were successfully developed. Having made a significant contribution to China's pharmaceutical history, North China Pharmaceutical Factory is well deserved as the "cradle of the pharmaceutical industry in the PRC." The completion and operation of the North China Pharmaceutical Factory not only enhanced the country's weak pharmaceutical production capacity but also boosted the process of industrial development in the PRC, enabling it to initially establish its own pharmaceutical industry foundation.

## The Elimination of Smallpox

In October 1950, the Government Administration Council of the Central People's Government of China issued the *Instruction on Launching the Autumn Smallpox Vaccination Campaign*, determined to promote nationwide vaccination. By 1952, over 500 million Chinese had been vaccinated throughout the country. Since 1961, with the recovery of the last case of smallpox, there has never been another case of smallpox in the Chinese Mainland.

## Successful Replantation of Severed Hands

On January 2, 1963, with the cooperation of vascular surgery expert Qian Yunqing, Chen Zhongwei, a doctor at Shanghai Sixth People's Hospital, successfully replanted Wang Cunbao's severed hand. Thereafter, China became the first country in the world to successfully replant a severed hand. Wang Cunbao's completely severed right hand regained normal function after the surgery, capable of carrying bags, writing, threading needles, heavy lifting, playing table tennis, etc. Chen Zhongwei pioneered the Chinese microsurgery technology, thus known as the "father of severed limb replantation" and "international pioneer of microsurgery" at home and abroad.

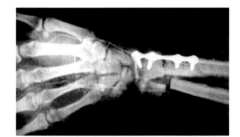

Shanghai Sixth People's Hospital successfully replanted Wang Cunbao's severed hand. The illustration is the radiography of the patient's artery on his hand and forearm two months after the surgery, which proves that the anastomotic vessels are unobstructed.

In 1963, two doctors, Qian Yunqing (left) and Chen Zhongwei (right) from Shanghai Sixth People's Hospital, were responsible for the replantation surgery of a severed hand. This operation had never been performed in the Chinese medical community. The illustration is the X-ray of the hand after the two doctors took over.

## Tang Feifan Separated Chlamydia Trachomatis

In 1955, Tang Feifan separated chlamydia trachomatis, the first scholar in the world to achieve that. Therefore, the trachoma virus is internationally named Tang Virus. The method he created is widely used. Subsequently, many similar pathogens were separated, and chlamydia, a special microorganism between bacteria and viruses, was discovered. Tang Feifan is the only Chinese microbiologist who has discovered important pathogens and opened up a research field so far. The confirmation of the trachoma pathogen has greatly reduced the incidence of trachoma worldwide. In 1982, at the International Congress of Ophthalmology held in Paris, the International Organization for the Prevention and Treatment of Trachoma awarded him the Gold Trachoma Medal in recognition of his outstanding contributions.

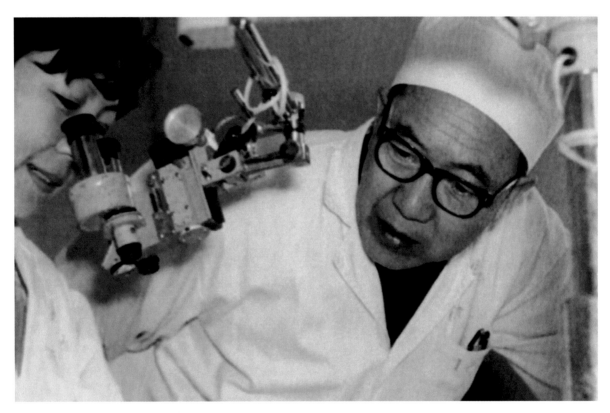

Wu Mengchao (1922–2021), a famous expert in hepatobiliary surgery, member of the Academic Divisions of CAS (academician), pioneer, and one of the main founders of the Chinese liver surgery department, known as the "Father of Chinese Hepatobiliary Surgery." He was the first doctor in China to win the highest national science and technology award. When he was young, he had a grand aspiration: "clear the notorious name of China being a liver-cancer country."

## A Hepatobiliary Surgery Miracle

The liver is the organ with the richest blood supply in the human body. It undertakes dozens of functions, such as material metabolism, digestion, storage, detoxification, blood coagulation, etc.

Today, medicine has created an artificial heart, artificial lung, and artificial kidney, but nothing can completely replace the complex function of the liver. The main treatment for liver cancer is surgical resection. In the 1950s, hepatobiliary surgery was regarded as a "forbidden zone" in the world and was at the exploration stage. At that time, there was no independent hepatobiliary surgery department in China, and liver surgery was still a blank page. Liver surgery often leads to death due to massive hemorrhage, and there were few successful operations in the world.

Wu Mengchao was a pioneer and one of the main founders of hepatobiliary surgery department in China. In the 1950s, he first proposed the new "five-lobe and four-segment" theory for Chinese liver anatomy; in the early 1960s, he invented liver resection in normothermic hepatic intermittent ischemia at room temperature and took the lead in breaking through the forbidden zone of the middle lobe of the liver, successfully performing the first complete resection of the middle liver lobe in the world, a breakthrough in the history of the world's major problems in liver surgery. Soon after, he performed three consecutive middle-liver lobectomies, all of which succeeded, which indicates the maturity of the liver surgery technology system created by Wu Mengchao. This drove the general development of liver cancer resection in China in the 1960s, decreasing the mortality of liver surgery in China from 33% in the 1950s to 4.83% in the 1960s and 1970s. The "five-lobe and four-segment" liver anatomy theory that he proposed is still in use, greatly improving the success rate of liver surgeries.

## (6) Support the Comprehensive Development of Basic Scientific Research

In order to prevent the neglect of basic research, *Basic Science Research Plan* was especially supplemented in the *Twelve-Year Science and Technology Plan* to strengthen the study of mathematics, physics, chemistry, astronomy, biology, and geography.

Basic science research fundamentally aims to deeply understand natural phenomena, reveal natural laws, acquire new knowledge, new principles, and new methods, and cultivate high-quality, innovative talents. It is an important source of high-tech development, the cradle of cultivating innovative talents, the foundation of building advanced culture, and the internal power of future scientific and technological advancement. The *Basic Science Research Plan* proposes to accelerate the development of basic science and technical science and fill in important gaps in this regard; to adhere to the combination of serving the national goal and encouraging free exploration when developing basic science research; to follow the law of scientific development; to attach importance to the exploration spirit of scientists; to highlight the long-term value of science; to provide stable support; to make advance deployment; to make dynamic adjustment according to the new trend of scientific development.

During the implementation of the *Twelve-Year Science and Technology Plan*, basic science research in China has begun to combine with the research of technical science and applied science, becoming an indispensable part of solving many major problems in economic and national defense construction. And many basic science research achievements have received high praise from the international academic community.

## Hua Luogeng's Theory of Functions of a Complex Variable

Hua Luogeng was the founder of many aspects of research in China, such as analytic number theory, classical groups, matrix geometry, automorphic function theory, and the theory of functions of a complex variable, and was also an outstanding representative of world-famous Chinese mathematicians. His research of the theory of functions of a complex variable began in the 1940s. Functions with complex numbers as independent variables are called functions of a complex variable, and the theory thereof is the theory of functions of a complex variable. The analytic function is a kind of function with analytic properties in functions of a complex variable. The theory of functions of a complex variable mainly studies analytic functions in the complex number field, so it is also called the theory of analytic function.

Hua Luogeng (1910–1985), mathematician, member of the Academic Divisions of the CAS (academician).

Wu Wenjun (1919–2017), mathematician, member of the Academic Divisions of the CAS (academician).

## Wu Wenjun's Topology

Wu Wenjun had been engaged in the research of algebraic topology since the 1940s and had made a series of significant achievements. Among them, the most famous is the introduction of the Wu class and the establishment of the Wu formula, which has important applications. It is widely acknowledged in the mathematical world that Wu has played a role as a link between the past and the future in topology research. In the 1950s, Wu and several famous mathematicians of his time worked together to promote the vigorous development of topology, making it one of the mainstream topics of mathematical science. For that, he won the first prize in the State Natural Science Award in 1956.

## Zhang Yuzhe Observed Asteroids

In the late 1950s, Zhang Yuzhe made outstanding achievements in the research of asteroid photoelectric photometry. Under his leadership, Purple Hills Observatory has published dozens of light variation curves of asteroids, and some of them were published for the first time in the world. Due to the high quality of observation, many curve data are widely cited by foreign researchers. Many high-quality papers and academic works he published have become classic literature. In addition, Zhang Yuzhe and his Purple Hills Observatory also studied the motion and physical properties of asteroids, providing valuable information about their origin and evolution. Some asteroids with special orbits discovered by them will probably become natural space stations, serving as a springboard for humans to trek to distant space.

Zhang Yuzhe (1902–1986), astronomer, member of the Academic Divisions of the CAS (academician).

Academician Wang Ganchang, a famous nuclear physicist, and pioneer in the research of experimental nuclear physics, cosmic rays, and particle physics in China, holds a high reputation internationally. In his 70 years of scientific research career, he had always been active at the forefront of science, diligently, and made many remarkable scientific achievements. He was a senior academician of the CAS and the winner of the "Two Bombs and One Satellite (missiles, nuclear bombs, and the artificial satellite)" Meritorious Medal, setting a high example for later generations.

## Wang Ganchang Discovered Anti-sigma Minus Hyperon Particles

In the late 1950s, a piece of sensational news spread from the Joint Institute for Nuclear Research in Dubna, the Soviet Union—the research team directly led by Wang Ganchang, a Chinese physicist working there, discovered anti-sigma minus hyperon particles when conducting experiments on the 10-billion-electron-volt proton synchronous phase stabilized accelerator. This discovery caused a great sensation at that time. The journal *Nature* pointed out that the discovery of anti-sigma minus hyperon particles in experiments eliminates a blank spot in the image of the micro world. Newspapers around the world have published detailed reports on this discovery, making Wang Ganchang one of the keywords in the news introduction.

## Zhu Xi Has Bred a Parthenogenetic Toad

From 1951 to 1961, Zhu Xi established the experimental system of hormone-induced amphibian ovulation in vitro to study oocyte maturation, fertilization, and artificial parthenogenesis. He found that the secretion of the fallopian tube was the decisive substance for the fertilization of the toad oosphere. He put forward the "ternary theory of fertilization" for amphibians and bred the firstbatch of "toads without maternal grandfather" in the world. He discovered that low-temperature dormancy is an indispensable external condition for the maturation of the oosphere of Bufo gargarizans (Asiatic toad), and proposed that the fertilization of mature oosphere of cyprinid fish, like amphibians', be closely related to the normal development of embryos, which theoretically guides the artificial incubation of Chinese carp.

He also conducted mixed-sperm hybridization research on silkworms and found that excessive sperm can affect the heredity of offspring, thus providing a new method for silkworm breeding. The domestication and cultivation of the ricin silkworm led by him, having solved the problems of hatching, breeding, and overwintering, were promoted in over 20 provinces across the country, adding a raw material to the textile industry. He won the National Invention Award in 1954 and the first prize in the Science and Technology Progress Award of the CAS in 1989.

Zhu Xi (1900–1962), an experimental embryologist, cytologist, ichthyologist, biologist, and member of the academic divisions of the CAS, used 14 strains of silkworm as materials to conduct mixed-sperm hybridization experiment according to the physiological multi-sperm fertilization characteristics of silkworm eggs. The illustration shows the cocoon obtained by Zhu Xi (left) after observing the silkworm mixed-sperm hybridization experiment.

## Feng Kang Created Finite Element Method

In the late 1950s and early 1960s, with the development of computers, scientific computation rose in the West. Feng Kang keenly realized that scientific development had entered a turning point, and China was given a rare opportunity. He led his research team to undertake a series of calculation tasks assigned by the state.

The opportunity to create the finite element method came from a national key task, namely the calculation of the design of Liujia Gorge Dam. Faced with such a specific practical problem, Feng Kang found a basic problem with sharp eyes. In 1965, he published the paper "Difference Scheme Based on Variational Principle" in *Applied Mathematics and Computational Mathematics*, proving the convergence and stability of the method under extremely extensive conditions while error estimates were made, thus establishing a strict mathematical theoretical basis for the finite element method, and providing a reliable theoretical guarantee for its practical application. The publication of this paper marks the independent creation of the finite element method by Chinese scholars. The creation of the finite element method is an important milestone in the development of computational mathematics.

Feng Kang (1920–1993), a mathematician, member of the Academic Divisions of the CAS (academician), founder and pioneer of computational mathematics in China.

Zou Chenglu (1923–2006, left), biochemist, member of the Academic Divisions of the CAS (academician).

## Zou Chenglu's Zou Plot

Zou Chenglu, one of the founders of modern Chinese biochemistry, has done significant pioneering work in the field of biochemistry. In his early years, he studied respiratory chain enzymes at Cambridge University under Professor Kelin, a famous biochemist. Zou returned to the field of enzymology in the early 1960s. In 1962, the quantitative relationship formula between the modification of protein functional groups and their biological activity Zou established was called Zou Formula, which was widely adopted by international peers; the mapping method he created to determine the number of necessary groups is called Zou Plot and has been included in textbooks and monographs. His achievements in quantitative research on the relationship between protein structure and function won him the first prize in the State Natural Science Award in 1987.

## Synthetic Insulin

Since 1958, the Shanghai Institute of Biochemistry of the CAS, the Shanghai Institute of Organic Chemistry of the CAS, and the Department of Biology at Peking University have joined forces. Headed by Niu Jingyi, a collaboration team has been assembled by Gong Yueting, Zou Chenglu, Du Yuhua, Ji Aixue, Xing Qiyi, Wang You, and Xu Jiecheng. On the basis of previous studies on insulin structure and peptide chain

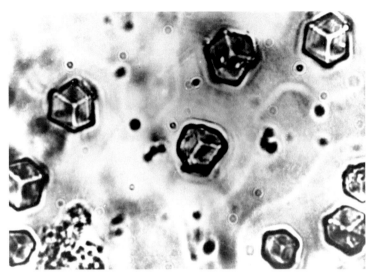

A crystalline form of synthetic insulin.

synthesis methods, they have begun to explore chemical methods to synthesize insulin. After a thorough study, they established a procedure for synthesizing bovine insulin.

The synthetic insulin in China has rapidly occupied a leading position in the world under a relatively weaker chemical basis of polypeptides. The illustration shows the researchers synthesizing B chain peptide and A chain peptide.

## Li Siguang Proposed the Theory of "Terrestrial Facies of Petroleum"

In 1959, geologist Li Siguang and others proposed the theory of "terrestrial facies of petroleum." Li clearly pointed out that "the key to finding oil is neither 'marine' nor 'terrestrial,' but whether there are conditions for oil generation and storage and the correct understanding of structural laws." His view breaks the Western scholars' theory of "dearth of oil in China." He summarized the favorable conditions for oil generation as follows:

1. There must be relatively broad low-lying areas, which have long been submerged by shallow sea or large lakes.
2. Around these low-lying areas, there used to be a large amount of biological reproduction, and at the same time, there had to be a huge amount of microorganism reproduction in the water.
3. A suitable climate for the growth of living things.
4. A great deal of mud and sand frequently transported from the land to the shallow sea or the big lakes, so that a great number of organic substances and microorganisms from land that grow and die extremely fast in water are quickly buried to prevent them from rotting into gas and spread into the air. Li Siguang's scientific foresight was confirmed. Oil layers were discovered one after another, ushering in the rapid development of the Chinese oil industry and declaring the complete failure of the "dearth of oil in China" theory, which once again confirmed the correctness of Li's theory.

Li Siguang (1889–1971), a scientist, geologist, educator, social activist, and member of the Academic Divisions of the CAS (academician).

## (7) Improve Scientific and Technological Teams and Strengthen International Exchanges

The implementation of the *Twelve-Year Science and Technology Plan* has played a decisive role in the establishment and layout of Chinese scientific research institutions, in the adjustment of disciplines and specialties in colleges and universities, in the training direction and use of science and technical teams, in the system and methods of science and technical management, and in the formation of the Chinese science and technology system.

In order to improve some important and urgently needed scientific departments in China so that they have a smaller gap with or catch up with the advanced international standards and meet the needs of national construction within 12 years, we must strive for international help while we should mainly rely on our own strength to develop science and technology. To this end, China has made full use of the strength of international cooperation in science and technology to vigorously conduct scientific and technological exchanges with other countries so that we can master the existing international advanced scientific and technological achievements in the shortest time, perform creative scientific research, rapidly improve the country's scientific and technological level, further enrich the world's scientific asset, and promote the common prosperity and development of science in all countries.

### Strengthen the Construction of Colleges and Universities

In order to strengthen the development of colleges and universities in inland areas and change the situation of their over-concentration, the central government decided to move some colleges and universities or departments in coastal areas inward and build or expand some colleges and universities based on certain inward moving majors.

In September 1958, in order to adapt to the needs of science and technology, national defense construction, and national economic development, and to cultivate the state's much-needed cutting-edge science and technical talents, the central government decided to have the CAS found a university in Beijing. Guo Moruo, the incumbent president of the CAS, was its first president. This is the University of Science and Technology of China (USTC). Compared with Peking University and Tsinghua University, USTC was a rising star. Its great number of outstanding students has become the backbone of scientific research institutions and universities.

### Establishment of the China Association for Science and Technology (CAST)

From September 18 to 25, 1958, the All-China Federation of Natural Science Societies and the All-China Association for Science Popularization jointly held a national congress in Beijing. Nie Rongzhen delivered an important speech on behalf of the Central Committee of the CPC and the State Council. The meeting approved the *Resolution on the Establishment of the China Association for Science and Technology*. Li Siguang was elected as the chairman of the first session of the National Committee, and Liang Xi, Hou Debang, Zhu Kezhen, Wu Youxun, Ding Xilin, Mao Yisheng,

Wan Yi, Fan Changjiang, Ding Ying, and Huang Jiasi as ten deputy chairmen. Yan Jici, Chen Jizu, Zhou Peiyuan, Tu Changwang, Xia Kangnong, and Nie Chunrong were elected as secretaries of the Secretariat.

The CAST is a mass organization of Chinese scientists and engineers, a people's organization under the leadership of the CPC, a link and bridge between the CPC and the government to connect scientists and technicians, and an important force for the country to promote its development of science and technology. Since its establishment, the CAST has adhered to the idea of revolving around the center and serving the bigger picture. It has always taken strengthening the tie between the CPC and the government and scientists and technicians as its fundamental responsibility, taken dedicated service to scientists and technicians as its fundamental task, and taken the satisfaction of scientists and technicians as the main criterion of its performance. It has made fruitful achievements in promoting the prosperity, development, and popularization of science and technology, in promoting the growth and improvement of science and technical talents, in promoting the combination of science and technology and economy, and in building a "home of scientists and technicians," which has been highly praised by the CPC and the people and won loud applause from the society.

The First National Congress of the CAST.

USTC was founded in Beijing in 1958 and moved to Hefei, Anhui Province in 1970. Known as the country's "cradle of science and technical talents," it holds a high reputation at home and abroad. The illustration shows Guo Moruo with students from USTC.

## Establishment of Beijing Center of World Federation of Scientific Workers (WFSW)

On September 25, 1963, the Beijing Center of the WFSW was officially established in Beijing. Biquard, Secretary General of the WFSW, and representatives of 21 countries from Europe, Africa, Oceania, and Latin America, as well as thousands of Chinese scientists, were invited to attend the inaugural meeting of the Beijing Center of the WFSW. The meeting was hosted by Zhou Peiyuan, the vice president of the WFSW and the vice president of the CAST. In his opening speech, he called the establishment of the Beijing Center of the WFSW "a major event in the life of the majority of scientific workers." Premier Zhou Enlai and Vice Premier Nie Rongzhen of the State Council met with representatives of various countries who attended this ceremony in Beijing. The WFSW was not only the first multilateral international science and technology organization the CAST joined but also an important platform for China to break the long blockade and communicate with the international science and technology community.

Representatives who attended the 1964 Beijing Science Symposium were visiting and investigating during the symposium.

## The Successful Symposiums

From August 21 to 31, 1964, the Beijing Science Symposium was held in Beijing. Three hundred sixty-seven scientists and government officials from 44 countries and regions in Asia, Africa, Oceania, and Latin America attended the event. Representatives from various countries included experts in both natural and social sciences. The findings and experiences of scientific research were changed, and issues of common concern to all, such as striving for and maintaining national independence, developing the national economy, culture, and science, and promoting scientific and

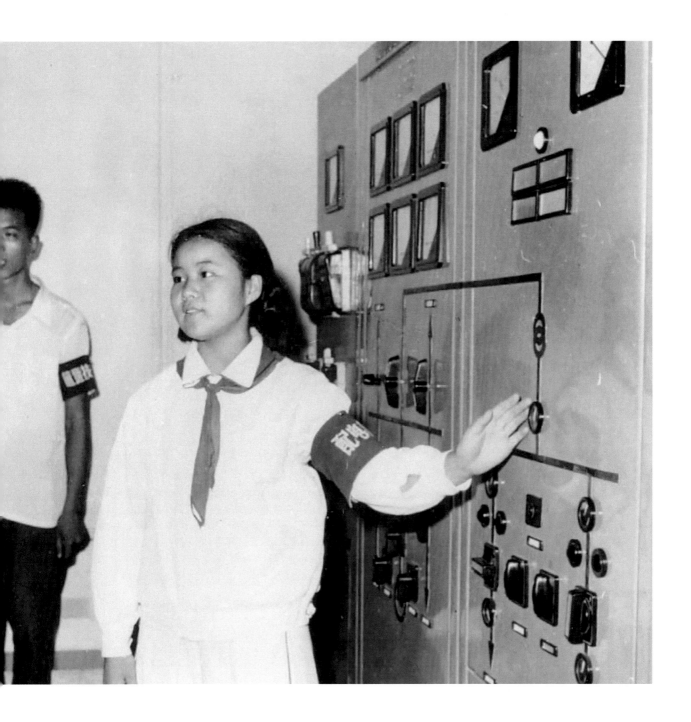

technological cooperation among countries, were discussed. During the symposium, the CPC and state leaders Mao Zedong, Liu Shaoqi, Zhu De, Zhou Enlai, Deng Xiaoping, Peng Zhen, Chen Yi, Nie Rongzhen, Tan Zhenlin, Lu Dingyi, Luo Ruiqing, Lin Feng, Yang Shangkun, Ye Jianying, Guo Moruo, Bao Erhan, and Zhang Zhizhong warmly received the delegates. The meeting had a significant and far-reaching impact on the further development of the scientific cause of countries on the four continents and the world.

From August 21 to 31, 1964, the Beijing Science Symposium was held in Beijing.

From July 23 to 31, 1966, the Summer Physics Symposium was held in Beijing. One hundred forty-four representatives from 33 countries in Asia, Africa, Latin America, Europe, and Oceania and one regional academic organization attended the event. Premier Zhou Enlai sent a congratulatory telegram. The Chinese delegation to the 1966 Summer Physics Symposium was composed of 36 people. Attending physicists from various countries submitted 99 academic papers in the fields of elementary particles, nuclear physics, solid-state physics, etc. The holding of this summer physics symposium carried forward the spirit of democratic consultation and active cooperation and enhanced the solidarity and friendship among scientists from four continents.

# Self-Reliance and Catching Up
# Implementation of the *Ten-Year Science and Technology Plan*

In the 1960s, the situation at home and abroad has undergone significant changes. At that time, the Soviet Union withdrew all researchers; in the turmoil of the "Anti-Rightist" and "Great Leap Forward" in China, the enthusiasm for scientific research was frustrated. In the winter of 1960, the Central Committee of the CPC issued the four-word policy of "adjustment, consolidation, enrichment, and improvement," requiring adjustments to be made in all walks of life. Under this circumstance, the *National Science and Technology Development Plan for 1963 to 1972* (referred to as the *Ten-Year Science and Technology Plan*) was proposed, which was the second science and technology development plan formulated based on the fact that the main tasks in the *Twelve-Year Science and Technology Plan* had been completed. The policy reads, "be self-reliant and catch up." It strives to build China into a powerful socialist country with modern industry, modern agriculture, modern science and technology, and modern national defense within a short historical period through diligent efforts; it emphasizes that "modernization of science and technology is the key to the modernization of agriculture, industry, and national defense." Despite the impact of the "Cultural Revolution," the *Ten-Year Science and Technology Plan* has made many gratifying achievements.

## (1) Achievements in Resource Survey

One of the objectives of the *Ten-Year Science and Technology Plan* is to strengthen the comprehensive survey of resources in China, strengthen the research on the protection and comprehensive utilization of resources, and provide the necessary resource basis for national construction. To this end, relevant scholars have conducted extensive investigations in the Yellow River basin, the Yangtze River basin, and the Huang-Huai-Hai Plain drafted a governance and development plan, and implemented the survey and exploitation of oil and gas resources in North China, Panjin, Jianghan, Central Plains, among other regions.

The oceangoing scientific survey vessels Xiangyanghong 5 and Xiangyanghong 11.

## First Scientific Survey of the Pacific Ocean

In July 1976, Xiangyanghong 5 and Xiangyanghong 11, Chinese oceangoing scientific research vessels, successfully performed China's first scientific survey of the Pacific Ocean and obtained a great amount of first-hand, multidisciplinary information.

# (2)  Achievements in Agricultural Science and Technology

Agriculture is the foundation of the national economy. The Central Committee of the CPC proposed to gradually realize the reform of agricultural technology on the basis of finishing the socialist reform of agriculture. From 1963 to 1972, the primary task of agricultural science and technology in China was to provide sufficient and accurate scientific and technological basis for completing this great and arduous historical task in a fast, efficient, and economical manner.

## Rational Utilization of Land Resources

Regarding the conditions for developing Chinese agricultural production at that time, some basic facts are listed as follows: there was not sufficient arable land, with an average of only two and a half mu of arable land per person (1 mu ≈ 667 square meters), and neither was there many wastelands suitable for farming. Of the 1.6 billion mu of arable land, about 2/3 was better land, while 1/3 had low yield. The level of mechanization, electrification, and chemistry of agriculture was relatively low, and many areas still needed further water conservation. In addition to 1.6 billion mu of arable land, the vast grasslands, hills, mountains, and waters in China were used properly to develop agriculture, forestry, animal husbandry, sideline production, and fishery on a large scale.

China has accumulated some experience and made some scientific research achievements in mastering modern production conditions, such as the use of pesticides, fertilizers, and agricultural machinery. Therefore, it was required to combine single-subject research with comprehensive research in agricultural research, summarize and improve farmers' production experience and the Chinese agricultural heritage, combine them with the development of modern science and technology, and combine scientific research with popularization. During the implementation of the *Ten-Year Science and Technology Plan*, many research and experimental projects have been completed, including the national arable land soil survey, soil improvement, rational fertilization, pest control, soil improvement, and cultivation techniques, sand control, alkali control, etc.

## Magical Oriental Rice of Yuan Longping

Yuan Longping took the lead in conducting research on the utilization of rice heterosis in 1964. He led a team to first discover male sterile plants, pointed out that rice has heterosis, and proposed the idea of utilizing heterosis through sterile lines, maintainer lines, and restorer lines. In 1972, the first rice male sterile line Erjiunan 1A and its corresponding maintainer line Erjiunan 1B were bred in China. In 1973, Nanyou 2, the first strong hybrid rice combination, was bred. In 1975, together with the members of the cooperation group, they overcame the technical difficulties in seed production, making China the first country in the world to successfully apply rice heterosis in production. Yuan Longping led his scientific team to give the world a powerful weapon to fight hunger. Therefore, China's hybrid rice is called the Magical Oriental Magic in the world.

Yuan Longping (1930–2021), academician of the CAE, known as the "father of hybrid rice" in the world.

Red Flag Canal.

## Construction of the Red Flag Canal

The Red Flag Canal is located in the north of Linzhou under the jurisdiction of Anyang City, Henan Province, where Henan, Shanxi, and Hebei provinces meet. The canal is composed of a trunk canal, three main canals, and hundreds of branch canals, with a total length of 2,000 kilometers. The total trunk canal is 70.6 km long, 4.3 meters high, and 8 meters wide, and has a water diversion volume is 20 m³/s. On April 5, 1965, the trunk canal of the Red Flag Canal was opened to water; in July 1969, the supporting construction of the trunk canal, branch canal, and head ditches was finished.

# (3) Achievements in Industrial Science and Technology

Industry plays a leading role in the national economy. There was a considerable gap between the Chinese industrial production level at that time and the advanced international level, which was mainly reflected in industrial technology. To seize the opportunity to improve the technical level of basic industries within ten years, we must establish new industrial sectors to raise the Chinese industrial development level to the international level in the 1960s. Only in this way could we ensure that in 20 to 25 years, we could basically realize the technical reform of agriculture, adapt to the requirements of the modernization of national defense, accelerate the development of industry, provide modern scientific research instruments and materials, and enable China's science and technology to approach and catch up with the advanced international level.

In order to achieve the industrial science and technology goals in the *Ten-Year Science and Technology Plan*, scientists have been "self-reliant and hard-working," making rich scientific and technological achievements, such as the design and construction of Panzhihua Iron and Steel Base, the Second Automobile Works, Chengdu–Kunming Railway, 10,000-ton oceangoing ships, enormous coal mines, large hydropower stations, and thermal power stations, etc., and the heavy machinery plants' manufacture of complete sets of equipment required by factories, mines, and railways.

## Second Automobile Works

In 1964, the Central Committee of the CPC included the construction of the country's Second Automobile Works (SAW) into the key project of the Third Five-Year Plan. Premier Zhou Enlai, on behalf of the Central Committee of the CPC, approved that the site of the SAW be located in Shiyan, Yun County, Hubei Province. On April 1, 1967, the commencement ceremony was held in Daluzigou, and large-scale construction began in 1969, which set off the climax of the construction of the SAW.

Shiyan City, Hubei Province, is the headquarters of Dongfeng Motor Group (formerly the SAW). There are more than 200 local industrial enterprises supporting Dongfeng Motor Group in the city, with a strong comprehensive supporting capacity.

# Dongfeng 10,000-Ton Freighter

Dongfeng 10,000-ton freighter was built by Jiangnan Shipyard. Under the policy of "self-reliant and catch up," building a 10,000-ton oceangoing freighter independently developed by China was one of the key projects of the *Ten-Year Science and Technology Plan*. Dongfeng was the first 10,000-ton freighter designed and built by China. Chinese started to design it in early 1958 and finished the entire construction design drawings in only three and a half months, creating a record for the high-speed design of large ships. On April 15, 1959, the cargo ship was launched with a berth period of 49 days. It reflected the ship design and manufacturing level and supported the ship production capacity of China at that time, laying a foundation for the country to build large ships above 10,000 tons in large quantities.

After the launch of Dongfeng, it happened to be a difficult three years when China was faced with a severe international environment. Most of the supporting equipment had to be developed by China itself. The installation project was at a standstill. The hull was moored along the Huangpu River for several years. It was not until the end of 1965 that the Dongfeng freighter was declared complete after the internal installation was finished and verified. On December 31, 1965, Dongfeng was officially delivered.

The successful construction of Dongfeng 10,000-ton freighter marks a new step for the Chinese shipbuilding industry.

## Chengdu–Kunming Railway Completed and Opened to Traffic

Chengdu–Kunming Railway is 1,091 km long from Chengdu South Railway Station to Kunming. It was started in 1958, opened to traffic in July 1970, and delivered for operation in December 1970. The route passes through the Daliang Mountains and the Xiaoliang Mountains, and the "One Line Sky" Canyon, which is two to three hundred meters deep. From Jinkou River to Aidai, there are 44 tunnels on the 58-km route. The 120 km section from Ganluo to Xide circles around the mountain four times for 50 km, crosses the Niuri River 13 times and has 66 km of tunnels and 10 km of bridges. After Xide, it crosses the Anning River eight times and the Jinsha River at Sanduizi. The Jinsha River valley is a famous earthquake zone in a fault zone. The route coils the mountains in the valley three times, crosses the Longchuan River 47 times, and stretches south to Kunming.

The earthwork of the Chengdu–Kunming Railway was nearly 100 million cubic meters, with 427 tunnels and 345 kilometers of extension, and 991 bridges and 106 kilometers of extension. The total extension of bridges and tunnels accounts for 41% of the total line length. 41 of the 122 stations along the line are located on bridges or in tunnels due to terrain restrictions. This railway is the skeleton of the road network in Southwest China, which is of great significance to the development of resources in that region, to the acceleration of national economic construction, and to the strengthening of national unity and the consolidation of national defense.

Since its operation, Chengdu-Kunming Railway, the main transportation artery in southwest China, has become the lifeline of the developing southwest economy and the link between the border and other regions in China. The illustration shows the railway under construction.

The Chengdu-Kunming Railway, built between Sichuan Province and Yunnan Province with complex terrain, is a miracle because of the precipitousness of the terrain it sits on.

## (4) Achievements in Medical Science

The general goal of medicine is to make remarkable achievements in protecting and improving people's health, preventing and controlling major diseases, and birth control, among other important scientific and technological issues. We must effectively solve the key scientific and technological problems in the prevention and treatment of serious diseases that affect the national economic construction and threaten people's health, so as to control and eliminate these diseases. Major achievements have been made in the theory of clinical medicine, prevention, and basic medicine and the application of some new medical technologies, and high-level medical research centers have been built. We have made contributions to summing up the clinical experience and the research work of traditional Chinese medicine (TCM) and acupuncture; in terms of using modern science to collate and study the rich Chinese medical heritage, a more complete and effective method has been formed, in terms of research on drugs, antibiotics, biological products, and medical equipment, scientific basis has been provided to improve quality and increase new varieties so that drugs and medical equipment are basically self-sufficient.

After the founding of the PRC, TB prevention and control centers were established in many regions of China, which provided favorable conditions for rural TB prevention. BCG vaccination has been conducted to prevent children from TB, and the number of vaccinations has increased year by year.

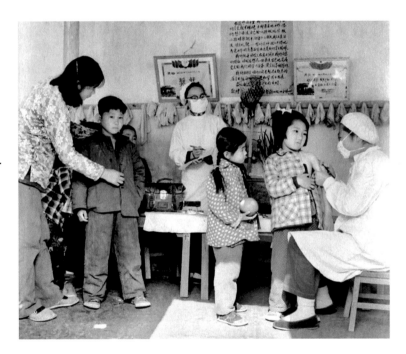

## Prevention and Treatment of Tuberculosis (TB)

TB is an ancient disease and one of the most important infectious diseases in the world. Back then, TB in China had six characteristics: first, a great number of people were infected; second, a great number of patients showed symptoms; third, there were many new patients; fourth, mortality was high; fifth, most patients being from rural areas, about 80% of TB patients in the country were concentrated in rural areas; sixth, many patients had drug-resistance, and this problem continued to deteriorate. Since the founding of the PRC, under the leadership of governments at all levels, a number of policies or measures have been introduced to improve the detection rate of TB patients, such as expanding the scope of free treatment, strengthening the registration and management of TB patients in medical institutions, and establishing a supervision and management system for TB patients, which has made great contributions to ensuring people's health.

The illustration shows senior doctors with years of medical experience and young doctors summarizing their bone setting experience.

## Summarize and Develop TCM

TCM is the summary of the extremely rich experience of the Chinese people's long-term struggle against diseases. It has a history of thousands of years. As an important part of the excellent Chinese culture, it is a science that studies human physiology, pathology, and the diagnosis and prevention of diseases. It is closely related to China's human geography and traditional academic ideas. China is a country with a vast territory and an enormous population. In many places, especially in the vast populous rural areas, people are still accustomed to using traditional methods to prevent and treat diseases. TCM has a certain potential in prevention, health care, health preservation, and rehabilitation. It consists of three parts: Chinese medicine, ethnic medicine, and folk herbal medicine.

Guo Shikui (1915–1981), an expert in internal medicine of TCM, devoted his whole life to the research on the prevention and treatment of coronary heart disease with TCM, developed the theory of promoting blood circulation to remove blood stasis, and worked with other people to create famous prescriptions such as coronary heart disease prescription II, chest painkiller pill and chest painkiller aerosol.

As time rolls forward, the application of modern science and technology in the field of TCM keeps expanding, and many new treatment methods and research findings have been discovered, rejuvenating the ancient TCM.

## Eliminate Schistosomiasis

Schistosomiasis was an epidemic in China, and it posed a great danger to people. The disease spread across 12 provinces and municipalities including Jiangsu and Zhejiang, with over 10 million people infected, and over 100 million under the threat of an infection. Schistosomiasis seriously harmed agricultural production and people's health in endemic areas, making its eradication an important political task at that time. In 1957, the State Council issued a directive on its eradication. Given that in some areas where schistosomiasis was spreading, there were other serious diseases as well, people's committees at various levels in areas where schistosomiasis was rampant to prevent and control the schistosomiasis, prepared conditions and gradually combined them with the prevention and control of other serious

Xie Yuyuan (1924–2021), pharmaceutical chemist, member of the Academic Divisions of the CAS (academician). The illustration shows that he was performing vacuum distillation on a newly discovered drug to treat schistosomiasis when he was an assistant researcher at the Institute of Medicine of the Shanghai Academy of Sciences.

diseases. In ethnic minority areas where schistosomiasis was endemic, when prevention and control were arranged and promoted, their life, production habits, and religious customs were fully taken into account, and publicity and education were carried out patiently to steadily prevent and control. Also, financial and technical support was provided.

The illustration shows medical staff in Taicang, Jiangsu Province, investigating schistosomiasis in rural areas.

## Treatment and Prevention of Leprosy

In January 1956, after the *National Agricultural Development Program (Draft)* issued by the Central Committee of the CPC clarified the task that leprosy be actively prevented and treated, the Institute of Dermatology of the Chinese Academy of Medical Sciences has repeatedly undertaken this national task of leprosy prevention and treatment. Leprosy is a chronic contagious disease caused by Leprosy bacilli, which mainly damages human skin and nerves. If not treated, it can cause progressive and permanent damage to the patient's skin, nerves, limbs, and eyes. This disease has a long epidemic history and is widely distributed, bringing serious disasters to people in epidemic areas. In order to control and eliminate it, the nation adhered to the policy of "prevention first," implemented the guiding principle of "active prevention and infection control," and executed the practice of "investigation, isolation, and treatment at the same time," actively finding and controlling the source of disease, and cut off the spread; meanwhile, the immunity of the surrounding natural population was improved by offering BCG vaccination to the children in the epidemic area, the family members of the patients, and the close contacts with negative leprosy and tuberculin reactions, or by giving them effective chemical drugs for preventive treatment.

Liu Wuchu (1924–), an expert in leprosy. He tried his best to relieve the suffering of the patients and gave all his efforts and talents to this end.

Tu Youyou (1930–), a pharmacist, has been engaged in the research of a combination of TCM and Western medicine for years and has made remarkable achievements. She led research teams to develop new antimalarial drugs, artemisinin and dihydroartemisinin, and won the Nobel Prize in Physiology or Medicine in 2015.

## The Internationally Acclaimed Innovative Drug—Artemisinin

In 1969, the Chinese Academy of TCM accepted the research task of antimalarial drugs, and Tu Youyou was appointed leader of the science team. However, due to various reasons, it was difficult for the research to take place. In 1971, Premier Zhou Enlai gave important instructions on this matter at the anti-malaria conference held in Guangzhou so that Tu Youyou could lead the research team to resume the research. As she searched through the medical literature of TCM, she noticed that Ge Hong, a famous doctor in the Eastern Jin Dynasty, wrote in the *Zhou Hou Bei Ji Fang* (handbook of prescription for emergency) that "steep a handful of sweet wormwood in water, and drinking this juice can cure chronic malaria." Following this clue, the research team improved the extraction method, and the obtained extract of Artemisia annua significantly increased its potency against malaria in mice; after that, in order to ensure the safety of the medication, she also tried it herself. In 1972, the research team isolated effective antimalarial monomers from Artemisia annua, which had a 100% inhibition rate against malaria parasites in mice and monkeys, and was named artemisinin. In the same year, Tu went to Changjiang, Hainan Province, to test the medicine, and 30 cases of artemisinin against malaria succeeded. As of 1978, a total of 2,099 cases of malaria were clinically cured, making artemisinin an eye-catching new antimalarial drug.

## (5) Achievements in Technical Science

Technical science, at its stage of rapid development, has become an important part of the modern science and technology system. The close connection of basic science, technical science, and engineering technology matters a great deal to the implementation of the principle of integrating theory with practice and to the improvement of science and technology. The task of all disciplines of technical science is to focus on studying scientific theories common to all sectors of industrial production and engineering technology so as to solve various problems of industrial production and engineering technology. On the one hand, we should comprehensively make use of the research findings of basic science; on the other hand, we should summarize the practical experience of production, combine the findings and experience, and develop a system theory. The development goal of technical science is to closely coordinate with the needs of national defense and economic construction, study and solve key problems in technical science, vigorously cultivate science and technical talents, develop modern experimental technology, and strive to approach or catch up with the advanced international level in some important fields.

### The First Experimental Man-Made Earth Satellite of China

On March 3, 1971, China successfully launched an experimental man-made earth satellite, which weighed 221 kg. Its operating orbit was 266 km from the perigee, 1,826 km from the apogee. The included angle of its orbital plane and the equatorial plane of the earth was 69.9°, and it took 106 minutes to circle the earth. It successfully transmitted back various scientific experimental data to the ground at the frequencies of 20,009 MHz and 19,995 MHz. The satellite was equipped with cosmic rays, X-rays, a high magnetic field, and ex-orbital heat flux detectors, enabling China to obtain space physics data via satellite for the first time.

### Laser-Controlled Nuclear Fusion

Nuclear fusion can release tremendous energy. To harness the tremendous energy generated by artificial nuclear fusion to serve mankind, we must make nuclear fusion under our control, which is controlled nuclear fusion. The realization of controlled nuclear fusion has an extremely promising future, but there are also many unimaginable difficulties ahead. Regardless, people have achieved satisfying progress. Scientists have devised many ingenious methods, such as using powerful magnetic fields to constrain reactions, using powerful lasers to heat atoms, etc. The research on controlled nuclear fusion in China began in the mid-1950s. The Southwest Institute of Physics of the Nuclear Industry (established in 1965) and the Institute of Plasma Physics of the CAS (established in 1978) were two institutions that specialized in the magnetic confinement of nuclear fusion research in China. Laboratories at the University of Science and Technology of China, Tsinghua University, Huazhong University of Science and Technology, and Beijing University of Science and Technology were also conducting relevant research work.

On March 3, 1971, China launched its first scientific experiment man-made earth satellite.

In 1975, in order to verify the ability of the laser to compress target materials, the Shanghai Institute of Optics and Precision Machinery established a six-channel high-power laser target-shooting experimental device with an output power of 100 gigawatts. This illustration shows the staff adjusting the laser fusion target room.

# (6) Achievements in Basic Science

The development of basic science has had a profound impact on the development of industry, agriculture, medicine, and military science and technology. Many major technological innovations and the emergence of new production technology departments are inseparable from new achievements in modern basic scientific research. They are often the fruits of the direct application of basic scientific research achievements. The level of basic science research is the level of the entire natural science research. To vigorously develop basic science research has become an important science and technology policy of countries with advanced modern science and technology.

Basic science mainly aims to accelerate the advancement of basic science and technical science, enrich the reserve of scientific theories, strengthen the accumulation of scientific investigation and experimental data, and establish and strengthen important and weak departments. We should effectively cooperate to solve the major scientific and technological problems in China's socialist construction, especially to make contributions to tackling key agricultural problems, developing cutting-edge technology, and making important achievements in some major scientific theoretical issues. Meanwhile, we should actively cultivate talents, establish and enrich research centers and experimental bases in a planned manner, and form the modern Chinese basic research system.

## Chen Jingrun's Goldbach Conjecture

Chen Jingrun had been aspiring to study the Goldbach conjecture since middle school. And he devoted his whole life to studying it. Chen adopted the sieve method to solve the problem, which involved extensive, complicated calculations. Concentrated and undistracted, he led a hermit life in a dormitory only six square meters big. The desk, the floor, the bed, and the wooden box were all piled with his scratch paper. Sacks of scratch paper on which he made the calculations were stuffed under the bed. After long unremitting efforts to explore, Chen Jingrun finally finished his paper and climbed up the "1+2" staircase. Previously, when five foreign mathematicians proved the "1+3" theory, they had the help of a large computer, while Chen Jingrun proved "1+2," that is, any sufficiently large even number can be expressed as the sum of two numbers whose prime number and prime factor do not exceed 2, with an old pen. This finding is called the Chen Theorem. So far, the Goldbach conjecture was only one last step away from being proved. Chen published the Goldbach conjecture in 1966. His achievements in studying it and other number theory problems are still far ahead in the world. Chen himself is also known as "the No.1 Man of the Goldbach conjecture."

## Important Achievements in Function Research

Yang Le and Zhang Guanghou, admitted to the Department of Mathematics of Peking University at the same time, were in the same class. In 1962, after graduation, they were both admitted to the postgraduate program of the Institute of Mathematics of the CAS. In the 1970s, Zhang Guanghou and Yang Le cooperated and co-founded the specific connection between the two main concepts of "deficient value" and "singular direction" in the theory of function value distribution for the first time, which was named Zhang-Yang Theorem by the mathematical circle.

Chen Jingrun (1933–1996), mathematician, member of the Academic Divisions of the CAS (academician).

## Zhu Kezhen's Achievements in the Study of the Paleoclimate of China

Zhu Kezhen was the founder of modern geography and meteorology in China. In 1961, he wrote "The Fluctuation of World Climate in Historical Times," and in 1972, he published academic papers such as "Preliminary Research on Climate Change in China in the Past 5,000 Years." The former paper, based on the geographical phenomena described in relevant documents and materials such as the attenuation of Arctic ice, the northward movement of the southern boundary of the Soviet Union's permafrost zone, the recession of the world's alpine glaciers, and the rise of the sea level, proved that the global climate had gradually warmed in the 20th century. From there, he traced back to the historical period, investigating the Quaternary global climate and the changes and fluctuations of floods, drought, and cold and warm weather in various countries, and found that the cold period in the lower reaches of the Yangtze River in the second half of the 17th century was consistent with the Little Ice Age in Western Europe.

Therefore, he pointed out that the change in solar radiation intensity might be an important factor in climate fluctuation. This provides new arguments for the study of historical climate. It is the crystallization of his decades of in-depth study of historical climate and a major academic achievement that stunned both at home and abroad. Making full use of the records in ancient Chinese documents and local chronicles, as well as the archaeological findings, phenological observation, and instrument records, he conducted research to find thwe truth and reached convincing conclusions.

Zhu Kezhen (1890–1974), a geographer, meteorologist, founder of Chinese meteorology, and member of the Academic Divisions of the CAS (academician).

# The Glory of "Two Bombs and One Satellite"

After the founding of the PRC, a great number of outstanding scientific works, including many scientists who have made outstanding achievements abroad, have devoted themselves to the sacred and great cause of building the PRC with great enthusiasm. Under the circumstances of weak national economic and technological foundation and poor working conditions at that time, self-reliant and diligent, they broke through in cutting-edge technologies such as the atomic bomb, hydrogen bomb, and artificial earth satellite with less investment and shorter time. What remarkable achievements!

For China, the "two bombs and one satellite" spirit is the attitude and process of making scientific and technological miracles; it is the embodiment of patriotism, collectivism, socialist spirit, and scientific spirit; it is a new precious spiritual wealth created by the Chinese people for the nation in the 20th century. We will continue to carry forward this great spirit and make it a great driving force for the people of all ethnic groups to stride ahead on the road of modernization.

## Beginning of the Aerospace Undertakings

On October 8, 1956, the Fifth Research Institute of the Ministry of National Defense, the first missile research institution in China, was officially established. This establishment was highly valued by Chairman Mao Zedong and other national leaders. It marks the official beginning of the Chinese aerospace undertakings.

## Construction and Start of Launch Center

In October 1958, China built its first space launch site on the Gobi Desert to the north of Jiuquan, Gansu Province. The area is flat and sparsely populated, with an inland and desert climate. It has sunny days all year round and long sunshine, which can provide a favorable natural environment for space launches. Jiuquan Satellite Launch Center is one of the launch test bases for scientific satellites, technical test satellites, and launch vehicles, and the earliest and largest integrated missile and satellite launch center in China.

Taiyuan Satellite Launch Center, founded in 1967, sits in the plateau area in the northwest of Taiyuan, Shanxi Province. In the temperate zone, the winter there is long, and there is no summer, meaning spring and autumn are connected. With only 90-day frost free, it is one of the test bases in China to test satellites, application satellites, and carrier rocket launches, and is mainly responsible for launching spacecraft in solar synchronous orbit and polar orbit. On December 18, 1968, the first medium-range carrier rocket designed and manufactured by China successfully took off there.

Xichang Satellite Launch Center was built in 1970 and completed in 1983. It is located in Sichuan Liangshan Yi Autonomous Prefecture in the hinterland of Daliang Mountain Canyon, 65 kilometers northwest of Xichang City. The area has a subtropical climate and ground wind that is gentle and moderate throughout the year. The best launch season is from October to May. It is mainly used to launch geosynchronous orbit satellites and is also the launch site of the lunar exploration project.

## Atomic and Hydrogen Bombs

It was an inevitable decision for China under specific historical conditions to develop nuclear weapons. In the early 1950s, the newly founded PRC was still under the threat of war, including the threat of nuclear weapons. The grim reality made Chinese leaders realize that to survive and develop, and we must have our own nuclear weapons and forge our own swords and shield. On October 16, 1964, the country's first atomic bomb successfully exploded. On June 17, 1967, China successfully conducted its first hydrogen bomb test, breaking the superpowers' nuclear monopoly and nuclear blackmail, and contributing to maintaining world peace.

## The First Artificial Earth Satellite

On April 24, 1970, China's first man-made earth satellite, Dongfanghong-I, was successfully launched, which is of epoch-making significance in its space history. China has therefore become the fifth country after the Soviet Union, the United States, France, and Japan to independently launch satellites. Dongfanghong-I artificial earth satellite was launched at Jiuquan Satellite Launch Site with the Long March 1 carrier rocket developed by China itself. This satellite is a spherical polyhedron with a diameter of about one meter and weighs 173 kg, heavier than the sum of the first man-made satellites of the Soviet Union, the United States, France, and Japan. The perigee and apogee of its orbit are 439 km and 2,388 km, respectively. The included angle between its orbital plane and the earth's equatorial plane is 68.5°. The time to circle the earth is 114 minutes. The Long March 1 carrier rocket, which sent the satellite into space, is a three-stage solid hybrid rocket that employs liquid and solid rocket motors. It is about 30 meters long and has a takeoff weight of 81.6 tons. The successful launch of Dongfanghong-I has laid a solid foundation for the development of Chinese space technology, driven the rise of the Chinese space industry, kept Chinese space technology in step with the world's space technology frontier, and marked the country's entry into the space age.

Xichang Satellite Launch Center.

The first hydrogen bomb was successfully detonated in China on June 17, 1967.

## Winner of "Two Bombs and One Satellite" Meritorious Medal

"Two Bombs and One Satellite" is a symbol of the great achievements of the PRC and national pride. In 1999, when people of all ethnic groups across the country celebrated the 50th anniversary of the PRC, the Central Committee of the CPC, the State Council, and the Central Military Commission solemnly commended those who had made outstanding contributions to the cause of "Two Bombs and One Satellite."

Winners of the "Two Bombs and One Satellite" Meritorious Medal are listed as follows:

Yu Min, Wang Daheng, Wang Xiji, Zhu Guangya, Ren Xinmin, Sun Jiadong, Yang Jiachi, Wu Ziliang, Chen Fangyun, Chen Nengkuan, Zhou Guangzhao, Qian Xuesen, Huang Weilu, Tu Shoue, Peng Huanwu, Cheng Kaijia

Posthumously:

Wang Ganchang, Deng Jiaxian, Zhao Jiuzhang, Yao Tongbin, Qian Ji, Qian Sanqiang, Guo Yonghuai

On April 24, 1970, China successfully launched its first man-made earth satellite.

# The Spring of Science

III

From March 18 to 31, 1978, the Central Committee of the CPC and the State Council held a grand National Science Conference in Beijing. Comrade Deng Xiaoping emphasized the basic Marxist view that science and technology are productive forces, highlighted the important strategic position of science and technology in economic and social development, pinpointed that the key to modernization is the modernization of science and technology; affirmed the dominant role of scientific and technical workers in scientific and technical activities, pointed out that intellectuals are part of the working class, and stressed that knowledge and talents should be respected. These assertions have clarified the major theoretical rights and wrongs that constrain the development of science and technology, broken through the shackles that have long been on the intellectuals, and laid the ideological and theoretical foundation for the basic guiding principles and policies of scientific and technological advancement in China in the new period, and greatly encouraged the innovation enthusiasm of scientific and technical workers across the country. Henceforth, it has opened the prelude to the reform of the Chinese scientific and technological system, marking the arrival of the spring of science in China.

The Third Plenary Session of the Eleventh Central Committee of the CPC, held in Beijing in December 1978, was a meeting of far-reaching significance in the CPC's history since the PRC's founding. It fundamentally broke through the heavy shackles of long-term "left" mistakes, corrected the guiding ideology of the CPC, and re-established the correct Marxist course of the CPC. After putting things right, China decided to shift the focus of the CPC's work to socialist modernization. After that, China began to reform the planned economic system. The guiding ideology of national economic and technological development has been established, and the scientific and technical goals and tasks have been clarified under the great call of "developing high-tech and implementing industrialization." In the end, the seeds sown in the spring of science have bloomed and fruited, and the development of science and technology has brought a thriving and earth-shaking glorious change to the great motherland.

# The Arrival of Spring of Science

This unprecedented reform and opening up in Chinese history have dramatically mobilized the enthusiasm of hundreds of millions of people, enabling China to successfully realize the great historical transition from a highly centralized planned economic system to a dynamic socialist market economic system, from closed and semi-closed to all-round opening. Facts have proved that reform and opening up is the critical choice to determine the fate of contemporary China and the only way to develop socialism with Chinese characteristics and achieve the great rejuvenation of the Chinese nation; reform and opening up is a powerful driving force to build socialism with Chinese characteristics. Only socialism can save the country, and only reform and opening up can develop China the country. We must develop socialism and Marxism.

The Central Committee of the CPC held the National Science Conference in Beijing on March 18, 1978, bringing together nearly 6,000 scientific and technical workers to denounce the Gang of Four, exchange experience, review achievements, and discuss plans for the first time since the founding of the PRC.

In the early spring of 1978, after a heavy snow, the weather warmed up day by day. A massive ideological liberation movement was brewing in the land of China, unstoppable. At the National Science Conference, people saw many familiar faces: Wang Daheng, Ma Dayou, Wang Ganchang, Ye Duzheng, Bei Shizhang, Zhu Guangya, Ren Xinmin, Yan Dongsheng, Yan Jici, Su Buqing, Yang Shixian, Yang Zhongjian, Wu Zhonghua, Wu Jichang, Wu Zhengyi, Shen Hong, Zhang Wei, Zhang Wenyou, Zhang Wenyu, Zhang Guangdou, Zhang Yuzhe, Lu Xiaopeng, Chen Jingrun, Mao Yisheng, Lin Qiaozhi, Jin Shanbao, Jiang Shengjie, Qian Sanqiang, Qian Xuesen, Gao Shiqi, Tang Aoqing, Huang Kun, Huang Bingwei, Huang Jiqing, Huang Jiasi, Liang Shoupan, Peng Shilu, Tong Dizhou... People began to break the forbidden zone, to think about some deep-seated problems that they dared not think about in the past, and actively seek new answers. Deng Xiaoping's speech at the National Science Conference was undoubtedly a magnificent declaration to liberate intellectuals and a banner calling for the dawn of a new era. Science and technology, a serious topic related to the fate and survival of the Chinese nation, has never been described in such a complete and systematic way nor so solemnly listed on the important agenda of the CPC and the country. In the passionate call for the spring of science, all scientists burst into tears, and the thunderous applause long echoed in the sky of the Great Hall of the People. The majority of scientific and technical workers felt from the bottom of their hearts that they were finally standing on a new starting point to serve the motherland with their talent.

## Deng Xiaoping's Southern Talks

At the beginning of 1992, Deng Xiaoping successively inspected Wuchang, Shenzhen, Zhuhai, and Shanghai, among other places, and delivered a series of important speeches, commonly known as "Southern Talks." In those speeches, he pointed out that there was no future without adhering to socialism, reform and opening up, developing the economy, and improving people's lives. Revolution liberates productive forces, and so does reform. We should be bolder in reform and opening up. We should dare to experiment. When we are certain, we should give it a valiant try. We should stick to what is right, correct what is wrong, and solve new problems as soon as possible. We should promote science because science is the only hope. We should grasp two key links simultaneously: reform, opening up, and cracking down on all kinds of criminal activities. His speeches unraveled the common doubts among people and affirmed the slogan "time is money, efficiency is life," which has been widely spreading throughout the country. He reiterated the necessity and importance of deepening reform and accelerating development, and, from China's reality, he stood at the height of the times, and profoundly summarized the experience and lessons of over ten years of reform and opening up, proposed new ideas on a series of major theoretical and practical issues, made breakthroughs, and significantly advanced the theory of building socialism with Chinese characteristics.

## Establishment of Special Economic Zones

On August 26, 1980, the Standing Committee of the National People's Congress formally adopted and promulgated the *Regulations of Guangdong Special Economic Zones*. China's special economic

The urban landscape of Shenzhen Special Zone.

zones were born. Shenzhen Special Economic Zone is one of the earliest experimental sites of reform and opening up that Comrade Deng Xiaoping opened up himself. At that time, the border town—Shenzhen Special Economic Zone he circled on the map has now become a highly modern city.

Special economic zones adopt more open and flexible special policies in their foreign economic activities than other domestic regions. They are areas where the Chinese government allows foreign enterprises or individuals, as well as overseas Chinese, Hong Kong, and Macao compatriots, to invest and enjoy specific policies. In special economic zones, foreign investors are provided with preferential conditions for the import and export of enterprise equipment, raw materials, components, corporate income tax rate, foreign exchange settlement and repatriation of profits, land use, residence, and entry and exit procedures of foreign people in business and their family members, etc. In other words, special economic zones are special economic regions where China adopts special policies and flexible measures to attract external funds, especially foreign funds, for development and construction. They are windows, pacesetters, and test grounds for the country's reform and opening up and modernization.

The special economic zones have shown strong vitality. First, their economy has continued to grow at high speed, and their development level has leaped to the forefront of the country; second, new breakthroughs have been made in reform, and the socialist market economic system has been basically established; third, remarkable achievements have been made in opening up, and a pattern of all-round opening up has taken shape; fourth, the ability of scientific and technological innovation has been significantly enhanced, and high-tech industries have flourished; fifth, people's living standards have been greatly improved, and the three civilizations have made common progress; sixth, urban construction, and management are becoming more and more modern, and the city has taken on a new look.

# The Rise of Rural Enterprises

Rural enterprises refer to all kinds of enterprises that are mainly invested by rural collective economic organizations or farmers and are founded in townships (including villages under their jurisdiction) to undertake the obligation of supporting agriculture. It is a general term for cooperative enterprises and individual enterprises of various forms, levels, categories, and channels in the rural areas of China, including township-owned enterprises, village-owned enterprises, farmer-owned cooperative enterprises, cooperative enterprises of other forms, and individual enterprises. There are many types of rural enterprises, including agriculture, industry, transportation, construction, commerce, catering, service, and repair, among other enterprises. Since the 1980s, rural enterprises in China have achieved rapid development. They have made full use of the natural and socio-economic resources in rural areas, marched toward the depth and breadth of production, promoted the prosperity of the rural economy and the improvement of people's material and cultural living standards, changed the single industrial structure, absorbed a significant number of rural surplus labor, enhanced the industrial layout, and gradually narrowed the gap between urban and rural areas and between workers and farmers. They are of great significance in establishing a new type of urban-rural relationship. The development of rural enterprises has become the only way for Chinese farmers to get rid of poverty and become better off. It is also an essential pillar of the national economy.

Zhuhai Hongwan Gas Turbine Power Plant was the first enterprise to enter Zhuhai Hongwan Development Zone. The completion of the power plant has positively contributed to Zhuhai's economic development and provided an essential power guarantee for the further development of the city's economy.

Huaxi Village founded Jiangsu Huaxi Industrial Corporation. At that time, there were 38 factories, including 5 Sino-foreign joint ventures. The life of farmers is improving day by day. The illustration shows a workshop at Huaxi Fine Wool Mill.

# Full Arrangement and Highlighted Key Points Implementation of the *Eight-Year Science and Technology Plan Outline*

In August 1977, at the symposium on science and education, Comrade Deng Xiaoping pointed out that China should start with science and education to catch up with the advanced international level. The current situation of science and education was poor, so we needed an organization to conduct unified planning, coordination, arrangement, guidance, and cooperation. Subsequently, various regions and departments began to start planned research preparation. In December 1977, the National Science and Technology Planning Conference in Beijing mobilized over 1,000 experts and scholars to participate in the research and formulation of the plan. In March 1978, the National Science Conference was grandly held in Beijing, and the *National Science and Technology Development Plan Outline (Draft) for 1978–1985* was reviewed and approved. In October of the same year, the Central Committee of the CPC formally forwarded the *Development Plan Outline (Draft) of National Science and Technology for 1978–1985* (referred to as the *Eight-Year Science and Technology Plan Outline*).

## (1) Scientific Plans and Related Work

### National Key Technologies R&D Program

The National Key Technologies R&D Program is a national mandatory plan. Its promulgation marks the emergence of China's comprehensive science and technology plan from scratch and is a milestone in developing the country's science and technology planning system. Since its implementation in 1983, the project has made significant progress in promoting agricultural development, upgrading traditional industries, developing major equipment, expanding new fields, and improving the ecological environment and health care level through science and technology. It has solved a number of complex technical problems involving national economic and social development and played an essential leading role in the technical development and structural adjustment of significant industries in China; meanwhile, a considerable number of scientific and technological talents have been cultivated, scientific research capability and technical foundation have been strengthened, and the overall level of China's scientific and technological work has been dramatically improved.

## Major Technical Equipment Development Program

The development program of major technical equipment, initiated in 1983, is a national mandatory science and technology program. It mainly supports the development of major technical equipment that significantly impacts national economic construction. In order to ensure the strategic focus of economic growth, it is necessary to determine the complete sets of technical equipment for the following ten major construction projects, organize the forces of all relevant parties, and introduce foreign advanced technologies for research, design, and manufacturing, mainly including the following:

1. Complete sets of equipment for large-scale open-pit mines with an annual output of ten million tons
2. Complete set of large thermal power generation equipment
3. Complete sets of equipment for the Three Gorges Hydropower Project
4. Complete sets of equipment for large nuclear power plants with a unit capacity of one million kilowatts
5. Complete set of ultra-high voltage AC and DC power transmission and transformation equipment
6. Complete sets of equipment for Phase II Project of the Baoshan Iron and Steel Complex
7. Complete sets of ethylene equipment with an annual output of 300,000 tons
8. Complete sets of large compound fertilizer equipment
9. Complete sets of large coal chemical industry
10. Complete sets of equipment for manufacturing large-scale integrated circuits

In November 1989, Heavy Ion Research Facility in Lanzhou (HIRFL) passed the complete acceptance of the National Appraisal Committee. HIRFL could accelerate total ions and provide a variety of stable nuclear and radioactive beams with a wide energy range and high quality for heavy ion physics and interdisciplinary research. The illustration shows the main accelerator of HIRFL.

National Key Project—Ningxia Chemical Plant.

## National Technology Development Plan

The National Technology Development Plan is one of the main bodies of the National Science and Technology Plan. It aimed to mobilize the enthusiasm of large and medium-sized enterprises for scientific and technological development, enhance their technological development capabilities, develop new products and technologies with high technical levels, significant economic benefits, and marketability, and promote the adjustment of product structure and industrial structure by using administrative and economic means such as planning, financing, and crediting.

## National Key Laboratory Construction Program

The national key laboratory construction plan was implemented in 1984. As an important part of the national scientific and technological innovation system, the National Key Laboratory is an essential base for the country to organize and apply high-level basic research, gather and cultivate outstanding scientists, and conduct high-level academic exchanges. Focused on the national development strategic objectives and faced with international competition, the National Key Laboratory conducts and applies basic research to enhance scientific and technological reserves and original innovation capabilities. It holds innovative ideas in the exploration of the scientific frontier; meets the needs of the national economy, social development, and national security, and achieves outstanding results in major key technology innovation and system integration; accumulates basic scientific data, materials, and information, and provides shared services and scientific basis for national macro decision-making.

## National Key Industrial Test Plan

The national key industrial test plan was implemented in 1984. Its main tasks were: to promote the transformation of scientific and technological achievements into productive forces as soon as possible, to expand the intermediate test results to a certain scale for testing, and to verify the feasibility and economic rationality of technology and equipment. The plan was both a national and local science and technology plan, meaning that the required funds were mainly composed of national, local, or departmental funds and self-raised funds by the units that undertook the projects.

## National Key New Technology Promotion Program

The national key new technology promotion program is a national guiding science and technology program, mainly oriented at enterprises. It primarily aimed to transform scientific and technological achievements into productive forces as soon as possible and serve economic construction. The program included new technologies, new processes, new materials, new designs, new equipment, and new agricultural varieties. New technologies were applied to transform traditional industries and improve the level of production technology and economic benefits.

## National Major Science Projects Plan

In 1983, the Chinese government, starting from the strategic goal of developing high-tech and taking a seat in the world high-tech arena, implemented the national major science projects plan. Major science projects refer to large-scale modern key instruments and equipment required for scientific studies. Because of their high construction level, great difficulty, and enormous investment, the national major science projects became an important symbol of a country's scientific and technological strength.

## (2) Implementation of Main Tasks of Scientific and Technological Research

The *Eight-Year Science and Technology Plan Outline* was the third long-term plan for the development of science and technology in China. It has made arrangements for research tasks in 27 fields, such as nature, agriculture, industry, national defense, and environmental protection, as well as basic disciplines and technical sciences. Among them, agriculture, energy, materials, electronic computers, lasers, space technology, high-energy physics, and genetic engineering, eight comprehensive science and technology fields that affect the overall situation, were given top priority. Major adjustments have been made in their implementation, and relevant policies have been formulated. Judging from the national strength at that time, the planned tasks and objectives obviously tended to excessive requirements and scale. With the shift of industrial focus, the science community further clarified that the strategic policy of science and technology should be oriented at economic construction.

On March 13, 1985, the Central Committee of the CPC promulgated the *Decision on the Reform of the Scientific and Technological System*, pointing out that the fundamental purpose of the reform of the scientific and technological system was to enable scientific and technological achievements to be rapidly and widely applied to production, so that the role of scientists and technicians can be fully played, thus greatly liberating scientific and technological productivity, and promoting economic and social development.

Since the promulgation of the *Decision on the Reform of the Scientific and Technological System*, some supporting reform measures have been gradually implemented and popularized, so the reform was widely carried out throughout the country.

Researchers actively conduct scientific experiments.

## The Advance of Agricultural Science and Technology Tasks

In accordance with the principle of "taking grain as the key link and developing in an all-round manner," a comprehensive survey of agricultural, forestry, animal husbandry, sideline, and fishery resources was conducted to provide a scientific basis for rational zoning and development and utilization. This agricultural principle was comprehensively implemented to ensure high and stable agricultural production. Farming systems and cultivation techniques compatible with mechanization have been developed. The South to North Water Transfer Project and related scientific and technical problems have been solved. Major progress has been made in improving low-yield soil and controlling soil erosion and sandstorm drought. The high yield, high quality, and stress resistance of improved varieties have been comprehensively enhanced. Compound fertilizer has been developed, and scientific fertilization has been applied. Biological and chemical simulation of nitrogen fixation has been studied. The integrated control technology for crop diseases and pests should be developed as soon as possible. Scientific research in forestry, animal husbandry, and fishery should be strengthened, and high-quality and efficient machinery and tools for agriculture, forestry, animal husbandry, and fishery should be developed; a comprehensive scientific experimental base for agricultural modernization should be built; the research on fundamental theories of agricultural science should be strengthened.

### 1) Significant achievements in regional governance and comprehensive development

Since the Sixth Five-Year Plan, China has established a group of ecological agriculture comprehensive pilot zones in the main ecological areas such as the Huang-Huai-Hai Plain, the Sanjiang Plain, the Loess Plateau, the arid regions in the north and the red loess soil areas in the south, and has achieved a wealth of experimental findings. The application of these findings has also produced significant social and economic benefits.

### 2) Fruitful results from the breeding of improved crop varieties

In terms of agricultural and livestock breeding, more than 30 new wheat varieties have been bred, with a regional experimental area of 40 million mu, accounting for 10% of the wheat sown nationwide and generally increasing production by 5%–10%; 40 new varieties of rice have been bred, and 50 million mu of rice has been promoted, with an average increase of about 50 kg per mu; 46 vegetable varieties have been bred and popularized in a wide area; the potato stem tip virus-free technology has been improved, and the technology of preventing potato degradation and yield reduction caused by virus invasion has been found. The average yield of potatoes per mu has been increased by 50%–100%, basically solving the technical problems of the national seed potato breeding system; four high-quality hybrid combinations and five fast-growing hybrid combinations have been screened from yellow feather broilers, which generally weighed 50% more and consumed 1/3 less feed when compared with local varieties.

At that time, China's crop germplasm resources were only second to the United States and the Soviet Union in the world. This advantage has provided a rich reserve resource for breeding improved crop varieties in China, promoted new breakthroughs in crop variety breeding technology, and achieved fruitful results in new variety breeding.

### 3) High and stable yield of Lu Cotton 1 in a large area

Lu Cotton 1 is a new variety bred by Shandong Cotton Research Institute, which has achieved a high and stable cotton yield. Cotton bollworm is a major pest in large-area production of cotton. Like cotton blight and verticillium wilt, it is a major factor leading to cotton yield reduction. Lu Cotton 1 organically combines the selection of transgenic Bt cotton with the utilization of cotton heterosis to improve cotton yield and insect resistance simultaneously so that the new varieties bred can achieve high and stable yields in production practice. In 1978, the Seed Bureau of the Ministry of Agriculture arranged 14 pilot projects in the two major cotton regions of the Yangtze River basin and the Yellow River basin, most of which received positive feedback.

Bagging in the process of sexual hybridization between wheat and rye. These newly cultivated fine varieties can increase the yield by 20%–40% compared with the original varieties, greatly increasing the unit yield.

Reclaim the Sanjiang Plain.

The Huang-Huai Hai-Plain is China's northern rice granary.

Hybrid rice seed production refers to the process of producing seeds by crossing sterile male lines with restorer lines as male parents. In 1973, the three-line hybrid rice was matched and put into field production.

### 4) Promising future of Li Denghai's hybrid corn

In the history of corn cultivation in today's world, there are two important figures: one is Wallace Carothers, a group leader at the DuPont Experimental Station laboratory and the holder of the high yield record of spring corn in the world; the other is Li Denghai, the creator of the world summer corn high yield record. In the field of breeding in China, there is the saying, "Yuan in the south and Li in the north." Yuan in the south refers to Yuan Longping, the father of hybrid rice, and Li in the north refers to Li Denghai, the founder of compact corn research, known as the "father of hybrid corn."

Li Denghai proposed the ideas that plant type and heterosis are complementary and that heterosis and population light energy are organically combined, which is a new breakthrough in breeding theory. The hybrid bred with his own "478" inbred line shows the ideal characteristics of compact corn with high light efficiency, the small angle between plant-type stems and leaves, and straight and upwelling leaves. Its leaf orientation value, extinction coefficient, population photosynthetic potential, and photosynthetic production rate, among other physiological indicators, are more reasonable, thus achieving the "four breakthroughs" in planting density, leaf area index, economic coefficient, and grain weight per plant under higher density. The planting density of his corn increased by 1,000–1,500 plants per mu on average.

Li Denghai (middle), a senior agronomist at Shandong Laizhou Corn Research Institute, is observing the growth of improved varieties of corn. This project has won the first prize in the National Spark Award. He demonstrated and promoted 25 new compact corn varieties, with a total promotion area of 520 million mu and an increase in grain production of 45 billion kg.

Lu Cotton 1 reaps a bumper harvest over large areas.

*5) Discovery of Jurassic angiosperm in Liaoning—archaefructus*

In 1990, Sun Ge, a researcher at Nanjing Institute of Geology and Paleontology of the CAS, and other scientific workers first discovered important early Cretaceous angiosperm fossils and pollen in situ in the Jixi area, Heilongjiang Province. In 1996, it was identified as archaefructus liaoningensis, which enabled the research on early angiosperms in Northeast China to take a great leap. On November 27, 1998, *Science* published a paper entitled "In Search of the First Flower: A Jurassic Angiosperm, Archaefructus, from Northeast China," written by Sun Ge and others as its cover article, revealing the true face of archaefructus liaoningensis to the world. Archaefructus liaoningensis belong to the family of archaefructus, including archaefructus liaoningensis and archaefructus Sinensis. They lived in the Mesozoic era 145 million years ago, 15 million years earlier than angiosperms found in the past, and are regarded as the earliest angiosperms by the international paleontological community. This provides strong evidence for the idea that flowering plants in the world originated in western Liaoning. Judging from the surface, the archaefructus liaoningensis fossils are well preserved, and their morphological characteristics are clearly visible.

Jurassic angiosperms found in Liaoning—archaefructus liaoningensis.

The illustration shows the fossil discoverer Sun Ge and his young assistant studying and observing the archaefructus liaoningensis fossils.

## Implementation of Energy Science and Technology

Various new exploration and development technologies are used to explore more oil and gas reserves, improve the level of drilling, oil and gas production, and develop petroleum geological theories. The key coal mines realize comprehensive mechanization, conduct research on coal gasification and liquefaction, develop large and efficient power stations and high-voltage power transmission networks, build atomic power stations, step up the research on the utilization of solar energy, geothermal energy, and biogas, actively explore new energy, and study the rational utilization of energy and energy saving technologies.

### *1) Petroleum exploration and development*

In 1981, implementing 100 million tons of contracted crude oil production and opening up, the whole petroleum industry played a decisive role in the petroleum industry's entry into a new stage of stable development. During this period, the national oilfield development adjustment has achieved remarkable results, making many achievements: the development technology of Daqing Oilfield has been improved, and the crude oil production has increased significantly; the exploration and development of Shengli Oilfield have opened up a new situation of oil development in eastern China; new high-yield oil and gas wells have been drilled in the western Qaidam region, a major breakthrough in Paleozoic marine oil and gas in China, becoming an epoch-making milestone in the history of oil and gas exploration in the country, and opening the prelude to oil and gas exploration in Tarim; the production of Liaohe Oilfield, Dagang Oilfield, Henan Oilfield, among other oilfields has also achieved breakthroughs.

## 2) Comprehensive mechanization of key coal mines

Comprehensive mechanization of the coal mining process refers to all production processes of coal mining in the stope face, such as coal breaking, coal loading, coal transportation, support, and roof management have been mechanized. In addition, chute transportation has also realized mechanization, giving full play to the efficiency of the comprehensively mechanized mining equipment. As comprehensively mechanized mining equipment continues to upgrade, the unit yield and efficiency of comprehensively mechanized mining faces are also rising.

## 3) Completion of Daya Bay Nuclear Power Station

Daya Bay Nuclear Power Station is located east of Shenzhen, 45 kilometers from Hong Kong. As China's first large-scale commercial nuclear power plant, it was put into commercial operation in 1994. It is the largest Sino-foreign joint venture in China, the second nuclear power plant built in mainland China, and the first nuclear power plant built in mainland China using foreign technology and capital. After that, Ling'ao Nuclear Power Station was built next to it. Together, they form a giant nuclear power base.

The completion of Daya Bay Nuclear Power Station marks that China's peaceful use of nuclear energy has reached the advanced international level. The entire power station has three major parts: nuclear island, conventional island, and supporting plant of the power station. The main equipment was imported from France and Britain. The main construction of the power station commenced in August 1987. The illustration shows the exterior of the power station.

(Above) Comprehensive mechanized coal mining process.

(Left) Due to the complex geological conditions of the Shengli Oilfield, the oil layers are mostly fractured oil and gas reservoirs. In order to improve the probability of oil production, directional deviated wells must be drilled.

## Material Science and Technology

As per the industrial policy of "taking steel as the key link," we have developed intensive mining technology, tackled the scientific and technological problems of red ore dressing and the metallogenic laws and prospecting methods of rich iron ores, solved the problem of comprehensive utilization of polymetallic ore resources, mastered a series of modern metallurgical new technologies, developed various special materials and composites required for the development of national defense industry and emerging technologies. At high speed, we have developed cement and light, high-strength, multi-function new building materials; studied the synthesis technology of organic raw materials based on oil, gas, and coal, developed new processes of synthetic materials, and strengthened the research of catalytic theory; carried out basic research of material science, developed new experimental testing technology, and gradually designed new materials per specified properties.

## Computer Science and Technology

Basic research in computer science and related disciplines has been conducted. The research on external equipment, software, and applied mathematics has been strengthened. We have solved the scientific and technological problems in the industrial production of large-scale integrated circuits, broken through the technological barrier of large-scale integrated circuits, and successfully developed 10-million-computations-per-second supercomputers and 100-million-computations-per-second supercomputers. We have formed the production capacity of computer series and vigorously promoted the application of computers and microcomputers. We have established a national public data transmission network and a number of computer networks and databases.

The high-quality, high-purity nano silicon nitride powder developed by the Anhui Institute of Optics and Precision has broad application prospects in machinery, energy, chemical industry, metallurgy, electronics, and national defense, among other fields.

The National Key Technologies R&D Program for the Seventh Five-Year Plan, "R&D of 1–1.5 μm CMOS Complete Process," undertaken by Tsinghua Microelectronics Research Institute, has made a breakthrough and developed a one-megabit Chinese character ROM for special large-scale integrated circuits urgently needed in China.

An ingot is produced.

The large-scale He-Ne laser holography system developed by the Beijing Oriental Science Instrument Factory of the Ministry of Aerospace Industry has passed the appraisal. The system has been applied to the nondestructive testing of various new satellites developed in China.

## Laser Science and Technology

We have quickly improved the level of common lasers. We have conducted basic research on lasers and made remarkable achievements in exploring new lasers, developing new laser bands, and using lasers to study material structures. As early as the 1960s, China established the world's first professional research institute in laser technology. In 1980, 1983, and 1986, China held three international laser conferences, which created conditions for academic exchanges between Chinese scholars and international laser researchers. After decades of endeavor, China's laser technology has built a strong scientific research force and strong technical foundation, trained a high-quality research team, and a great number of scientific and technological talents have been active at the forefront of laser research at home and abroad, harvesting fruitful results.

## Space Science and Technology

We have conducted research in space science, remote sensing technology, and satellite applications. We have developed a series of launch vehicles and developed and launched various scientific and application satellites for astronomy, communications, meteorology, navigation, broadcasting, and resource exploration. We have actively carried out studies on launching space laboratories and space probes and built a modern space research center and satellite application system. On November 26, 1975, China launched a recoverable remote sensing satellite for the first time, with a diameter of 2.2 meters and a height of 3.14 meters. The nature of this satellite is the same as that of the earth's resource satellite, except that it has a short working life. Its remote sensing instruments, such as cameras, can take lots of earth observation photos with the advantages of high resolution, small distortion, and moderate scale. It can be widely used in scientific research and industrial and agricultural production, including land survey, petroleum exploration, railway route selection, marine, and coastal mapping, cartography, target point positioning, geological survey, power station site selection, earthquake prediction, grassland and forest area survey, historical relics and archaeology, etc.

Since the establishment of the China Institute of Aerospace Medical Engineering on April 1, 1968, it has made outstanding achievements in the research of manned space medical engineering. Aerospace Life Assurance System Medical Engineering Research and Application won the first prize in the National Award for Science and Technology Progress in 1985. The illustration shows an astronaut performing a vestibular physiology test in the human swivel chair cabin.

On April 8, 1984, China launched the first experimental communication satellite, and on April 16, it was successfully positioned over the equator at 125° east longitude. It was put into use at the test stage and achieved good results.

FY-1 polar-orbiting meteorological satellite.

# High-Energy Physics

High-energy physics, also known as particle physics or elementary particle physics, is a branch of physics. It studies the structural properties of substances in the micro world deeper than the atomic nucleus, the phenomena of mutual transformation of these substances at very high energy, and the causes and laws of these phenomena. It is a basic discipline and one of the frontiers of contemporary physics. Particle physics, based on experiments, is developed on the close combination of experiments and theories.

At present, particle physics has gone deeper than the study of the properties of hadrons. Building a higher energy accelerator undoubtedly provides a more powerful means for experimental particle physics research. It is conducive to generating more new particles to clarify quarks and leptons' types, properties, and possible internal structures.

## *1) China's first proton linear accelerator*

On December 17, 1982, the first proton linear accelerator in China, built in the Institute of High Energy Physics of the CAS, extracted a proton beam with an energy of 10 million electron volts for the first time. The proton linear accelerator is a device that linearly accelerates protons. It consists of a high-frequency power supply, ion source, acceleration electrode, target chamber, vacuum system, etc. The acceleration electrodes, called drift tubes, are arranged in a straight line and alternately applied with high-frequency voltage to accelerate protons. Protons are sped up in the gap of the drift tubes. After entering the drift tube, they are protected from the influence of the deceleration electric field. Proton linear accelerator is widely used in industry and medicine, among other areas.

On April 26, 1989, China's first dedicated synchrotron radiation device was officially completed and commissioned. The main instruments of this device are an electron storage ring with an energy of 800 million electron volts and a linear accelerator with an energy of 200 million electron volts. 24–27 beam lines can be built in the synchrotron radiation zone. The illustration shows the electron linear accelerator in the synchrotron radiation device.

## 2) The Beijing Electron Positron Collider

In 1984, the Beijing Electron Positron Collider (BEPC) project, a national key project, broke ground. On October 16, 1988, the BEPC successfully achieved collision, another major breakthrough in the high-tech field in China after the explosion of atomic bombs, hydrogen bombs, and the launch of satellites. This was the first high-energy accelerator in China and a major science infrastructure for high-energy physics research. It is composed of a linear accelerator with a length of 202 meters, a transport line, a circular accelerator with a circumference of 240 meters (also known as the storage ring), a Beijing spectrometer with a height of 6 meters and a weight of 500 tons, and a synchrotron radiation experimental device around the storage ring, which looks like a giant badminton racket. The positive and negative electrons are accelerated to near the speed of light in the high vacuum tube and collide at the designated location. The particle characteristics generated by the collision are recorded by the large detector—Beijing Spectrometer. Through processing and analyzing these data, scientists can further understand the properties of particles, thus revealing the mysteries of the micro world.

The BEPC, China's first high-energy accelerator, successfully achieved collision.

In 1992, new data on the τ Particle mass was measured on the BEPC. In 1993, the first infrared free electron laser was created in Asia.

## Genetic Engineering

As genetics research deepens from the cell level to the molecular level, genetics has become the core discipline at the forefront of life science. The *Eight-Year Science and Technology Plan Outline* requires to establish and strengthen relevant laboratories, to develop basic research on genetic engineering, combining it with research on molecular biology, molecular genetics, and cell biology to make achievements in some important aspects of biological science that are close to or reach the advanced international level, to actively explore the possible ways of applying genetic engineering in the fermentation industry, agriculture, medicine, etc.

*1) Major breakthroughs in the research of synthetic ribonucleic acid (RNA)*

The research on synthetic RNA began in 1968. It was another important basic theoretical topic after China synthesized bovine insulin in the world for the first time. Chinese scientific researchers have cooperated with each other. After years of hard work, they first completed the preparation of raw nucleotides and the organic synthesis of small fragments of nucleotides. On this basis, they fully used the catalytic effect of enzymes. After trials and errors, they successively completed the synthesis of three fragments of 10 nucleotides, 12 nucleotides, and 19 nucleotides at the end of July 1979. After extensive experiments, scientists connected the three fragments and finally synthesized an RNA semi-molecule composed of 41 nucleotides. Like natural RNA, it has rare nucleotides. And the joints of the three fragments were verified correctly. This research was completed by the Shanghai

Institute of Biochemistry, the Institute of Organic Chemistry, the Institute of Cell Biology, the Institute of Biophysics of the CAS, the Department of Biology of Peking University, and relevant factories.

In September 1974, Chinese researchers adopted organic chemistry and enzymatic synthesis methods to connect nucleotides into eight small nucleotide fragments, which brought the synthesis of nucleotide fragments in China closer to the advanced international standard at that time; three years later, 16 nucleotides were synthesized.

*2) Xu Rigan led the breeding of test tube goats and test tube cows*

Since the 1970s, Xu Rigan had long been engaged in studying high technology of modern animal husbandry centering on the reproductive biology of livestock. In October 1983, during his further study in Japan, after over 400 experiments in the laboratory, he finally observed the whole process of goat IVF under the microscope. On the evening of March 9, 1984, Xu Rigan personally delivered the world's first test tube goat. After returning to China, he set up his own laboratory with the support of the Inner Mongolia Autonomous Region and has since conducted research on livestock biology and biotechnology focused on in vitro fertilization of cows and goats. This is a road that Chinese people have never walked, but with his perseverance and team, Xu has made one miracle after another in front of the world. In 1989, he and his team cultivated China's first batch of test. tube goats and test tube cows, lifting Chinese research in this field into the world's advanced ranks.

Xu Rigan (1940–2015), an expert in livestock reproductive biology and biotechnology, was the former vice president of the Chinese Academy of Engineering. The illustration shows that he was operating the microscope to observe animal embryos.

# (3) Achievements in Industrial Science and Technology

In terms of industry, we have mastered the digital seismic prospecting technology of oil, the construction technology of semi-submersible offshore oil drilling platforms, and the exploitation technology of gently inclined medium-thick mineral deposits; we have broken through the beneficiation technology of lean hematite, the blast furnace smelting technology of high titanium magnetite, the comprehensive utilization and recovery technology of vanadium and titanium, the extraction technology of rare earth elements and the mining, beneficiation, smelting and comprehensive recovery technology of copper, nickel and other symbiotic ores; have developed and put into use complete sets of spinning and post-treatment equipment with an annual output of 15,000 tons of polyester staple fiber; have constructed the large-scale water conservancy and hydropower project of Gezhouba Dam.

## The Construction of Shanghai Baosteel Group Corporation Commenced

On December 23, 1978, Shanghai Baosteel Group Corporation (from now on referred to as Baosteel) broke ground. This high-quality iron and steel base and the R&D base of new processes, new technologies, and new materials in the iron and steel industry, focused on the principal business of iron and steel, is a large-scale iron and steel complex with the highest degree of modernization,

Baosteel, based on its principal business of iron and steel, takes the road of diversified development, and has also made great progress in trade, finance, information, transportation, and construction, among other industries. The illustration shows its chemical equipment imported from the United States, Japan, and European countries.

the largest production capacity and the most complete varieties and specifications in China. It has gathered talents in engineering construction, equipment manufacturing, installation, and commissioning, among other aspects. As Baosteel's Phase I and Phase II projects have been put into operation, its construction has gone from the initial introduction of complete sets of equipment to the cooperative manufacturing of Phase II and Phase III projects, and finally, independent integration; the equipment manufacturing, installation, commissioning, and so on of Baosteel Engineering received instruction from foreign experts, and have gradually become able to manufacture and integrate domestically. To ensure the high quality of Baosteel's engineering and domestic equipment, Baosteel actively learns from the successful experience of construction supervision and adopts internationally accepted project management methods to control the whole process of its engineering and equipment manufacturing, installation, commissioning, etc.

## Panzhihua Iron and Steel Base

On November 27, 1978, the first large-scale iron and steel complex in China, Panzhihua Iron and Steel Base Phase I was completed and put into operation. Located in the southwest of Sichuan Province, its construction was started in 1965. By 1970, the blast furnace had cast iron. By 1975, the project's first phase had been basically completed and put into operation, gradually forming a total production capacity of 1.6 million to 1.7 million tons of iron, 1.5 million tons of steel, 1.25 million tons of bloom, and 900,000 to 1.1 million tons of steel. By 1980, the output of all main products had reached and exceeded the original design level. By 1985, the accumulated profits and taxes were equivalent to the state's total investment in the project's first phase.

Panzhihua Iron and Steel Base, located in Panzhihua City on the Jinsha River at the junction of Sichuan and Yunnan, enjoys the reputation of Jinsha Pearl. It is 749 kilometers from Chengdu in the north and 351 kilometers from Kunming in the south. The famous Chengdu–Kunming Railway runs through the city from north to south.

China's oil drilling and production equipment manufacturing industry has made great progress as the offshore oil industry develops. The illustration shows the Nanhai No.1 drilling platform testing in Beibu Gulf.

## Offshore Oil Drilling Platform

From June 25 to July 6, 1984, the first semi-submersible offshore oil drilling platform, Exploration No. 3, independently designed and built by China, successfully passed the test and was put into use in the East China Sea. The maximum drilling depth of this drilling platform can reach 6,000 meters, which meets the needs of offshore oil development on the Chinese Mainland shelf. Back then, only the United States, Japan, Britain, and Norway, among other countries with developed shipbuilding industries, were able to design and build such platforms independently. This Chinese semi-submersible platform mainly comprises a superstructure, submerged body, columns, and inclined struts. The submerged body includes shoe type, rectangular barge hull type, and strip pontoon type. Its appearance resembles a bottom-supported platform, and its superstructure is equipped with all drilling machinery, platform operating equipment, material reserves, and living facilities. It is a space box structure composed of a top plate, bottom plate, side wall, and several vertical and horizontal silo walls. It has high water tightness and can provide greater buoyancy. During operation, the submerged body is poured into the ballast water so that it can dive to a certain depth. It is positioned by the anchor cable or dynamic positioning. On a trail run, ballast water is discharged to make the submerged body float on the water's surface. When operating in shallow water, the submerged body can be placed on the seabed like a bottom-supported platform. It can work in the sea area with a depth of 10–600 meters, adapt to poor marine conditions, and exhibit good motion characteristics.

## Gezhouba Became Grid-Connected

In December 1988, the Gezhouba Water Control Project was completed. Gezhouba is located about 2,300 meters downstream of Nanjing, the outlet of the Three Gorges in Yichang, Hubei Province. After the Yangtze River leaves the Three Gorges, the water makes a sudden turn from the east to the south, and the river surface immediately widens from 390 meters to 2,200 meters at the dam. Due to sediment deposition, Gezhouba Island and Xiba Island are formed on the river surface, dividing the Yangtze River into the big, second, and third rivers. The big river is the main channel of the Yangtze River, and the second and third rivers dry up in the dry season. The Gezhouba Water Control Project spans the Yangtze River, Gezhouba, the Second River, Xiba, and the Third River. It is the first dam built on the long Yangtze River in China and an important part of the Three Gorges Water Control Project on the Yangtze River. The backwater of Gezhouba Reservoir is 110–180 kilometers.

Because the water level has been raised, 21 rapids and nine dangerous shoals in the Three Gorges have been submerged. Therefore, nine one-way channels and nine warping stations have been canceled, thus greatly improving the channel, enabling all kinds of ships in the waters below Badong County to sail unimpeded, and increasing the passenger and freight traffic volume of the Yangtze River. Gezhouba Water Control Project generates electricity and improves the channel of the Xiajiang River.

Its power generation capacity is enormous, reaching an annual output of 15.7 billion kilowatt-hours, thus saving 10.2 million tons of raw coal annually. It has played an important role in changing the energy structure of central China, reducing the pressure on the coal and oil supply, and improving the assurance of safe operation of central and eastern China power grids. The design level and construction technology of this vast water conservancy project, which is one of the few in the world, reflects the achievements of hydropower construction in China at that time, making it a milestone in the history of hydropower construction in China.

Gezhouba Dam is the first large-scale water conservancy project on the long Yangtze River and the stepped architecture of the Three Gorges Water Control Project, which was built later. It has greatly raised the Yangtze River's flood control capacity from managing a once-in-a-decade flood to a once-in-a-century flood.

Wang Xuan (1937–2006), academician of the CAS and the CAE. He insisted on taking the market as the guide and actively promoted the transformation of scientific and technological achievements into productivity. The Chinese character laser phototypesetting system he developed is one of the most comprehensive and successful examples of computer technology applications in China.

## (4) Achievements in Emerging Technologies and Basic Theoretical Research

In terms of emerging technology and basic theoretical research, we have successfully developed massive computers able to perform hundreds of millions of computations per second; solved the problems in the technology involved in 4,380 channel carrier communication of medium coaxial cable; conquered the radio measurement and control system of satellite carrier, experimental communication satellite, and microwave measurement and control system; launched the Long March 3 rocket; created the production technology of vitamin C two-step fermentation by using bioengineering technology; mastered the technology of hepatitis B virus core antigen and antigen preparation; developed new nonlinear laser materials—potassium phosphate and bismuth germanate crystals; overcame the hot forging technical difficulties of non-metallic artificial crystal materials and synthesized mica and RNA. These scientific and technological achievements have not only played a positive role in improving industrial and agricultural production but also have approached or reached the world's advanced level.

### Wang Xuan Led the Successful Development of the Chinese Character Laser Phototypesetting System

Wang Xuan is known as the "contemporary Bi Sheng (Bi Sheng was a Chinese artisan, engineer, and inventor of the world's first movable type technology, in the Song Dynasty)." The Chinese characters laser phototypesetting system he developed triggered a major technological revolution in the Chinese printing industry. The Huaguang and Founder electronic publishing systems, which he led to develop, once occupied 99% of the domestic newspaper market, 90% of the book (black and white) publishing market, and 80% of the overseas Chinese newspaper market, and entered the Japanese and Korean markets, obtaining huge economic and social benefits.

## Artificial Synthesis of Yeast Alanine Transfer RNA

The yeast alanine transfer RNA synthesized in China is the first RNA artificially synthesized in the world with the same chemical structure and complete biological activity as natural molecules. Since 1968, Chinese scientists have begun to study the artificial synthesis of yeast alanine transfer RNA. After thousands of trials and errors, two half molecules containing 35 and 41 nucleotides were respectively connected. The final synthesis was completed on November 20, 1981, and five repeated synthesis tests were carried out afterward, all of which succeeded. This was another feat for China in this field after the artificial synthesis of crystalline bovine insulin in 1965, marking another great step forward in exploring life science.

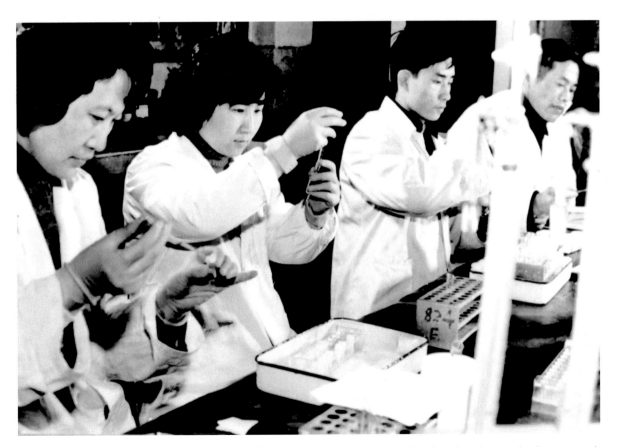

The researchers put the test tubes containing synthetic molecules, natural molecules, and other control tubes into the test box, and the data obtained showed that the artificial synthetic yeast alanine transfer RNA has the same biological activity as the natural molecules.

Established the Great Wall Station in Antarctica.

## The Great Wall Station in Antarctica

Scientists deem Antarctica as "the key to unlock the mysteries of the earth" and "the holy land of natural science experiments." Isolated, Antarctica is free from air pollution, which provides excellent conditions for observing celestial bodies; there are thousands of meteorites in Antarctica, making it a precious base for exploring the mysteries of outer space; Antarctica is one of the sources of the earth's atmospheric circulation, which has an important impact on global climate change; the extinct creatures in other parts of the earth six million years ago may be found in Antarctica. These findings may help us unravel the mystery of the origin of life on the planet and provide a scientific basis for further cracking the puzzle of evolution in the sea and on the land. The first China Antarctic Expedition team left Shanghai on November 20, 1984, and arrived at George Island in the South Shetland Islands of Antarctica on December 26. At 15:00 on the 30th, two landing boats, Great Wall 1 and Great Wall 2, carrying 54 expeditioners, boarded the south of Fildes Peninsula and hoisted the first five-star red flag there. At 10:00 on the 31st, on George Island, Antarctica, the cornerstone inscribed Great Wall Station in Antarctica from the motherland was erected on the land of Antarctica. On February 15, 1985, the expedition team announced to the world that the Great Wall Station in Antarctica was successfully completed, and a grand opening ceremony was held on George Island on February 20. In October, at the 13th Antarctic Treaty Consultative Parties Meeting held in Brussels, China formally became one consultative party because of the establishment of a long-term research station in Antarctica and its fruitful multidisciplinary research.

## Study on Cambrian Explosion

The Cambrian explosion is an outstanding case in paleontology and geology. Cambrian is a time in geological history. It was named after a hill in Britain from 545 million years ago to 495 million years ago. In July 1984, Hou Xianguang, a researcher at the Nanjing Institute of Geology and Paleontology, discovered the first Early Cambrian animal fossil Naraoia in Maotian Mountain, Chengjiang County, Yunnan Province. This discovery was purely accidental. Hou Xianguang intended to look for Bradoria in Chengjiang but accidentally stumbled upon the fossil of Naraoia (an early marine arthropod). Foreign scientists believe that the Naraoia is one of the earliest hard-bodied creatures. This is the first time it was discovered in the Asian continent with appendages. This discovery means evidence of the Cambrian explosion is right under our nose. Later, this day became the anniversary of the Chengjiang biota. German professor Adolf Seilacher, a world-renowned paleontologist, called the discovery of Chengjiang biota as shocking as a message from the aliens. *The New York Times* praised it as one of the most amazing discoveries of the 20th century.

The specimen of Cinderella eucalla, one of the fossils found in Chengjiang. It has a pair of giant eyes in front of the ventral surface of its head shell.

## The Yinhe Supercomputer

On December 6, 1983, China's first Yinhe supercomputer with over 100 million calculations per second was developed through the cooperation between the National University of Defense Technology and more than 20 units. It passed the state verification in Changsha. At that time, only a handful of countries in the world were able to develop such supercomputers. It realized the goal of a Chinese super-high-speed supercomputer put into use by 1985 proposed by the National

Science Conference two years ahead of schedule, enabling China to step into the ranks of countries capable of developing supercomputers, marking that its computer technology has entered a new stage. In November 1992, the Yinhe-II supercomputer was successfully developed and passed state verification. In June 1997, the Yinhe-III supercomputer system passed the state technical verification again. Its computing speed was 13 billion times per second, and its comprehensive processing capacity was over ten times that of the Yinhe-II, but its size was only 1/6 of that of the Yinhe-II. The comprehensive technology of the Yinhe-III supercomputer system reached internationally advanced standards at that time.

The trial calculation shows that extensive data in meteorology, petroleum, earthquake, nuclear energy, and aerospace, among other fields, can be processed at high speed on the Yinhe-II supercomputer. The illustration shows the developers of the supercomputer debugging the mainframe.

## "One Rocket & Three Satellites" in Orbit

On September 20, 1981, China successfully launched a group of three satellites, the Scientific Experiment Satellite 9 for Science Experiments, for the first time with only one rocket. This successful launch of "One Rocket & Three Satellites" marked a major breakthrough in China's aerospace industry. The three satellites are Practice 2, Practice 2A, and Practice 2B. This "One Rocket & Three Satellites" technology was the first time in China back then. In addition to its scientific significance, it reflected the military industry and aerospace technology level and caused a global sensation. After the three satellites entered orbit accurately, all systems functioned normally and continued to send various scientific detection and test data back to the ground.

Practice 2 Satellite.

# Adhere to the Strategic Policy of "Orienting at and Relying on"—Implementation of the *Fifteen-Year Science and Technology Plan* and the *Ten-Year Plan and the Outline of the Eighth Five-Year Plan*

In April 1981, the Central Committee of the CPC and the State Council instructed the National Science and Technology Commission to draft a scientific and technological development plan. More than 200 experts, having had a thorough discussion and listened to the opinion on the international trend of science and technology and analysis on their domestic experiences from famous experts from Germany, Japan, and the United States, among other countries, have formulated the *1986–2000 Science and Technology Plan* (also referred to as the *Fifteen-Year Science and Technology Plan*).

To implement the *Fifteen-Year Science and Technology Plan*, the National Planning Commission and the National Science and Technology Commission have prepared the *National Plan for Science and Technology Development during the Seventh Five-Year Plan Implementation*, including the *National Plan for Tackling Key Problems in Science and Technology during the Seventh Five-Year Plan Implementation*.

These plans have played an important role in mobilizing scientific and technological forces to serve the national economy. They have transformed science and technology into productive forces and closely oriented science and technology to economic construction. They strove to integrate with economic development, promote the improvement of scientific and technological awareness of the whole society, and have brought enormous economic and social benefits to the country.

Therefore, they have strongly confirmed that "science and technology are the first productive force," laid a solid foundation for the continued development of China's scientific and technological undertakings and enhanced the support system. It has been proved that during the Seventh Five-Year Plan period, centered on the primary task of combining science and technology with the economy, China's science and technology have undergone new changes by continuing to promote the transformation of technological achievements, reforming the allocation system for science and technology and other measures.

On the basis of a comprehensive analysis on the demand of China's economic and social development for scientific and technological advancement and on the trend of international economic and technological development, considering the scientific and technological strength, solid work foundation, problems and difficulties in front of us, and in accordance with the spirit of the Seventh Plenary Session of the Thirteenth Central Committee of the CPC and the goals set out in the country's *Ten-Year Plan for National Economic and Social Development and the Outline of the Eighth Five-Year Plan*, in March 1991, with the support and cooperation of all relevant departments, the National Science and Technology Commission began to formulate the *Ten-Year Plan for the Development of Science and Technology of the PRC from 1991 to 2000 and the Outline of the Eighth Five-Year Plan* (shortened as the *Ten-Year Plan and the Outline of the Eighth Five-Year Plan*), which is also the fifth time that China comprehensively formulated a long-term plan for the development of science and technology. The *Ten-Year Plan and the Outline of the Eighth Five-Year Plan* continued to adhere to the strategic principle of "orienting at and relying on" and further defined the goals and tasks of scientific and technological development in the next ten and five years, respectively.

## (1) National Science and Technology Plan Issued and Related Work during the Seventh Five-Year Plan Period

### The Spark Program

The Spark Program is the first program initiated in 1986 to promote rural economic development relying on science and technology under the approval of the Central Committee of the CPC and the State Council. It is an important part of China's national economy and science and technology development plan. The main task of the Spark Program is to earnestly implement the policy of the Central Committee of CPC and the State Council on vigorously strengthening agriculture and promoting the healthy development of rural enterprises, to guide the adjustment of agrarian industrial structure, to increase effective supply, and to encourage the development of agriculture through science and education. It has actively and practically promoted the transformation of the rural economic growth mode from extensive to intensive, relied on scientific and technological progress to improve labor productivity and economic efficiency, and guided farmers to change their traditional production and lifestyles. It has built a group of spark technology-intensive areas and regional pillar industries led by science and technology, promoted the scientific and technological progress of critical industries of rural enterprises, propelled the economic development of central and western regions, trained rural-applicable technology and management talents, and improved the overall quality of rural workers.

On March 2, 1986, Wang Ganchang, Yang Jiachi, Wang Daheng, Chen Fangyun (from the right), and other scientists jointly put forward the Proposal on Tracking and Studying the Development of Foreign Strategic High Technology to the Central Committee. Henceforth, China's high-tech research entered a new stage of planned and organized development on a national scale.

Science and technology revitalize agriculture. The Sanming Agricultural Correspondence University of Fujian Province specializes in agricultural technology education. Tu Jingchun (left), a teacher in the art department of Sanming Agricultural School, is a favorite person of villagers in Xishan Village, Fenggang Town, Shaxian County.

## Natural Science Foundation of China (NSFC)

The NSFC, based on the practice carrier of "effectively strengthening basic research, striving to improve the original innovation ability, and serving the construction of an innovative country," gives full play to the advantages of the NSFC system, promotes the balanced and coordinated development of disciplines, encourages scientists to conduct extensive exploration in the scientific front and national strategic needs, provides talent and project reserves for the implementation of major national science and technology projects and science and technology plans, and strongly supports the enhancement of national independent innovation capacity.

## National High-Tech R&D Project (Project 863)

Project 863 is a national high-tech R&D project organized and implemented by the Chinese government with important strategic significance for the country's long-term development in the critical period when the world's high-tech was booming and international competition was becoming increasingly fierce. It occupies an extremely important position in developing China's science and technology and shoulders the critical historical mission of developing high-tech and realizing industrialization. Following the spirit of the *Outline of the High-Tech R&D Project of the Central Committee of the CPC*, Project 863, based on the trend of global high-tech development and China's needs and practical possibilities, adhering to the policy of "limited goals, highlighted key points," has selected biotechnology, aerospace technology, information technology, laser technology, automation technology, and energy technology and new materials as the focus of China's high-tech R&D (marine technology was added in 1996). Its overall goal is to gather a small number of capable talents in selected high-tech fields, targeted at the world's highest level, to narrow the gap with developed countries, drive scientific and technological progress in related areas, create a new generation of high-level technical personnel, prepare for the formation of high-tech industries in the future, and make conditions for China's economic and social development to a higher level and for more robust national defense at the end of the 20th century, especially at the beginning of the 21st.

## Poverty Alleviation through Science and Technology

We have introduced many advanced and applicable scientific and technological achievements and means to poor areas, promoted and enhanced their self-development capabilities, and achieved sustainable and coordinated development of the economy, society, and environment in impoverished areas.

## Torch Program

The Torch Program was a guiding program for developing Chinese high-tech industries. It was approved by the Chinese government in August 1988 and organized and implemented by the

Ministry of Science and Technology of the PRC (formerly the State Scientific and Technological Commission of the PRC). The program aimed to implement the strategy of rejuvenating the country through science and education, execute the general policy of reform and opening up, give play to the advantages and potential of China's scientific and technological strength, and promote the commercialization of high-tech achievements, the industrialization of high-tech commodities, and the internationalization of high-tech industries under market guidance.

## National New Products Program

The national new product program was a policy support program launched by the Ministry of Science and Technology of the PRC in 1988. It aimed to guide and promote the scientific and technological progress of enterprises and scientific research institutions, improve their technical innovation ability, optimize the industrial structure and adjust the product structure, and accelerate the development and industrialization of high-tech products with strong economic competitiveness and large market share through independent domestic development and the digestion and absorption of foreign advanced technology.

The 1 MW power supply device that uses high-pulse to emit high-voltage pulse developed by the Institute of Plasma Physics, CAS.

Shandong University developed a green-light solid laser that can be used for information processing, laser printing, and compact disk.

## National Promotion Program of Scientific and Technological Achievements

The National Promotion Program of Scientific and Technological Achievements, based on the overall requirements of the national environmental construction of scientific and technological industrialization, adopts national policy guidance and measures to obtain bank loan support. It focuses on strengthening the promotion system and ecological construction and supports standard technologies that can enhance traditional industries and significantly impact the development of high-tech industries, as well as social public welfare technologies that have substantial economic, social, and ecological benefits. Through the joint organizations of states, ministries, and regions, it promotes scientific and technological achievements in a planned and focused manner to achieve economies of scale.

## National Soft Science Research Program

The National Soft Science Research Program is a major scientific and technological program that provides macro consulting services for national development. It was formulated to apply well, approve, and manage soft science research projects, lay a foundation for preparing the national soft science research plan, and promote the scientific and standardized management of soft science research.

## Military-to-Civilian Technology Development Plan

The military-to-civilian technology development plan is to actively support and guide enterprises and institutions of the national defense science and technology industry to develop high-tech products and industrialize, accelerate the establishment of innovative systems and mechanisms, promote the adjustment of civil product industry and product structure, and propel the development of national defense science and technology industry economy through the pull of special funds. The national defense science and technology industry has thoroughly implemented the strategic policy of "combining military with civilians and integrating military with civilians." While ensuring the completion of military research and production tasks, it has developed a great number of military-to-civilian technologies and products by taking advantage of military technology, talents, and equipment, thus promoting the economic growth of the national defense science and technology industry, driving the development of relevant industries of the national economy, and making significant contributions to economic construction.

Using the original scientific and technological achievements to develop civil products, Guizhou Fengguang Power Plant has built an advanced 75 mm integrated circuit production line to manufacture various integrated circuits and semiconductor separation devices.

## (2)  National Science Plan Issued during the Eighth Five-Year Plan Period and Related Work

### National Engineering (Technology) Research Center Program

The National Engineering (Technology) Research Center Program is an important part of building a social scientific and technological innovation system and the scientific and technological plan. In front of the general trend of global development and application of science and technology, in accordance with the needs of the development of the national economy and the socialist market economy, to accelerate the reform of the scientific and technological system, promote the transformation of scientific and technological achievements into actual productive forces, and enable China's economic construction to be timely adjust to the track of relying on scientific and technological progress and improving the quality of workers, we started with strengthening engineering research to master the common and key technologies required by enterprises, provide the engineering research environment and means of system integration in a targeted way, and promote the economic construction and social development towards relying on science and technology and taking an intension-type road.

### National Major Basic Research Project Plan (The Climb Plan)

With the rapid advancement of science and technology, basic research's great promotion and strategic influence on the national economy and social development have raised global attention. To reflect the guiding role of national goals in basic research, the National Major Basic Research Project Plan (The Climb Plan) was formulated in 1991 to organize major key projects that were relatively mature in basic research and that were comprehensive and stimulating for national economic and social development and scientific and technological progress, and implement them in the form of national mandatory plans. The Climb Plan has played a crucial role in strengthening and realizing the primary research objectives in China.

### Industrial Project Plan and Demonstration Program of National Major Scientific and Technological Achievements

To accelerate the transformation of major achievements, industrialize a batch of major scientific and technological achievements that have been obtained in line with the direction of industrial development, give full play to its advantages of high value-added benefits, provide demonstration and model projects for capital construction and technological transformation, and play a leading and promoting role in the adjustment of industrial structure, China has arranged a trial "special loan from the National Development Bank for industrialization of major scientific and technological achievements" in capital construction investment to support major, comprehensive and complete scientific and technological achievements industrialization projects that have a significant impact on the development of the national economy.

The Southwest Institute of Physics of the Nuclear Industry in Chengdu realized the operation of high constraint mode under the configuration of a diverter for the first time on the controlled nuclear fusion experimental device—China Circulator 2A. This is a significant milestone in China's magnetic confinement fusion experimental research history. Its comprehensive strength and level of magnetic confinement fusion energy development and research have been greatly improved.

## Productivity Promotion Center

The Productivity Promotion Center is a science and technology service institution that develops and disseminates advanced productive forces and assists small and medium-sized enterprises (SMEs) in technological innovation in the market economy. As an intermediary institution that is in line with international standards and provides social services for SMEs, it has made important contributions to improving the technological innovation capacity and market competitiveness of enterprises, especially SMEs', playing an irreplaceable role.

## Establishment of the China Academy of Engineering (CAE)

On June 3, 1994, the inaugural meeting of the CAE was held in Huairentang, Zhongnanhai, Beijing, giving birth to the first batch of Chinese academicians. After the founding of the PRC, the CPC and the government attached great importance to the development of engineering technology. As early as 1955, when the Academic Divisions of the CAS were founded, the Department of Technical Science was set up. Multiple Party and state leaders attended the inaugural meeting of the CAE and delivered important speeches. Then-Chairman Jiang Zemin personally signed, "Congratulations on the establishment of the CAE."

On June 8, 1994, the CAE elected its President Zhu Guangya (middle), and four vice presidents Lu Liangshu (first from the left), Zhu Gaofeng (second from the left), Shi Changxu (second from the right), Pan Jiazheng (first from the right).

The CAE is the highest honorary and consultative academic institution in the field of engineering technology in China and a public institution directly under the State Council. The CAE has established an academician system. As the highest academic title in engineering technology established by the state, a CAE member is elected from outstanding engineering science and technology experts who have made systematic and creative achievements and contributions in the field of engineering science and technology and is a lifelong honor. The CAE member election is held every two years.

## Young Scientists Forum by China Association for Science and Technology (CAST)

From June 12 to 13, 1995, the opening ceremony and the first Young Scientists Forum hosted by CAST were held in Beijing. Wu Jieping, Vice Chairman of the Standing Committee of the National People's Congress, Song Jian, State Councilor and Director of the State Scientific and Technological Commission, Zhu Guangya, Chairman of the CAST, Zhou Guangzhao, President of the CAS, and Wang Daheng, a famous photologist attended the event. Twenty-eight young scientists engaged in life science research, including Feng Changgen, one of the top ten outstanding youth in China, Bai Chunli, He Fuchu, and Dr. Ma Keping, participated in this first event and discussed some hot issues in life science. The young scientists, who attended the event, spoke highly of this high-end young

In June 1995, the Young Scientists Forum hosted by the CAST opened in Beijing.

scientist forum, believing that the CAST has created an excellent academic exchange atmosphere for young scientists to learn from each other. It has encouraged young scientists to move to the forefront of science and technology in the world, which is conducive to training and creating more cross-century outstanding scientific and technological talents and academic leaders. During the meeting, young scientists suggested that the forum follow the policy of contention of a hundred schools of thought, adhere to a scientific attitude of seeking truth from facts and a refined style of study, advocate academic democracy and academic freedom, and pay special attention to emerging disciplines, marginal and interdisciplinary disciplines in its topic selection. Since then, the Young Scientists Forum has been held all over the country. By the end of 2018, 370 sessions of the forum had been held.

# (3) Major Scientific and Technological Frontier R&D

During this period, major scientific and technological achievements, reaching or approaching the world's advanced level in some fields, have been promoted extensively, thus enhancing the technical level and economic benefits of traditional industries. The successful development of the Yinhe supercomputer, the successful launch of the underwater missile, the Long March 2 high-thrust strap-on rocket, the Asia-1 communication satellite, and so on indicate that China has made breakthroughs in high-energy physics, computer technology, carrier rocket technology, satellite communication technology, etc.

## Long March 2 High-Thrust Strap-On Rocket

In 1986, to meet the needs of the international satellite launch market and promote the further improvement of aerospace technology, China put the development of a high-thrust strap-on rocket on its agenda. The most important feature of the Long March 2E was the use of advanced strap-on technology, which greatly improved the rocket's carrying capacity and met the requirements for launching heavy low-orbit satellites at that time. The successful development of the Long March 2E carrier rocket has greatly enhanced the Chinese rocket's low orbit and geosynchronous transfer orbit carrying capacity and played a connecting role in developing China's high-thrust rocket. In 1995, Long March 2E had three international commercial launches in one year. On January 26, when launching the American-made Asia Pacific II communication satellite, the rocket exploded, destroying both the satellite and the rocket. This was a fiasco of the Long March 2. After a series of rectifications, it stepped out of the haze, succeeding again on November 28 and December 28 in accurately sending the two communication satellites, Asia-2 and Echo Star One, into their scheduled orbit. Consequently, China's Long March 2E high-thrust rocket has stepped onto the international satellite launch stage with its majestic posture.

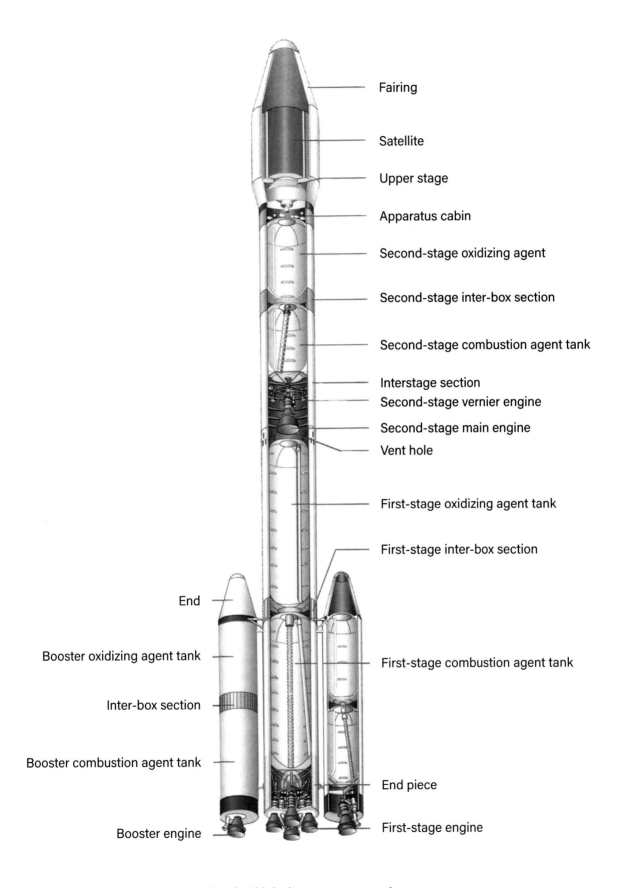

Fairing

Satellite

Upper stage

Apparatus cabin

Second-stage oxidizing agent

Second-stage inter-box section

Second-stage combustion agent tank

Interstage section
Second-stage vernier engine
Second-stage main engine
Vent hole

First-stage oxidizing agent tank

First-stage inter-box section

End

Booster oxidizing agent tank

Inter-box section

Booster combustion agent tank

First-stage combustion agent tank

End piece

Booster engine

First-stage engine

Long March 2 high-thrust strap-on rocket.

In 1988, China's nuclear submarine successfully launched missiles underwater.

Asia-1 communication satellite.

## Successful Launch of Underwater Missiles

In September 1988, China's nuclear submarine successfully launched missiles underwater. This marks a qualitative leap in the combat effectiveness of the People's Liberation Army Navy. The underwater ballistic missile launch test has figured out a path for China's nuclear submarines to evolve from an attacking type to a strategic type, thus enabling Chinese nuclear submarines to make a nuclear strike.

## Successful Launch of Asia-1

On April 7, 1990, China successfully launched the Asia-1 communication satellite with the Long March 2 strap-on rocket at the Xichang Satellite Launch Center. Hughes Aircraft of the United States, the manufacturer of the Asia-1 communication satellite, and experts from AsiaSat, together with Chinese aerospace experts, carried out this launch cooperation. Thousands of Han, Yi, and Tibetan people, among other ethnic groups, as well as more than 200 guests from 17 countries, Hong Kong and Taiwan, China, gathered at the launch site and witnessed the spectacle of launching foreign satellites via Chinese carrier rockets. This marks China's official entry into the international space launch market.

## The Chinese Volume of Human Genome Sketch Completed

On December 1, 1999, a research team composed of scientists from the United Kingdom, the United States, Japan, and other countries announced that the first pair of human chromosome genetic codes had been deciphered, another major breakthrough in human science. The Human Genome Project

(HGP) is a scientific feat that attracts global attention and transcends national boundaries and time. It has a great historical significance in promoting science and technology's development and benefitting humanity. On August 26, 2001, the Ministry of Science and Technology, together with the CAS and the NSFC, organized relevant experts to inspect and accept the Partial Sequencing Project of the HGP in China at the National Human Genome Northern Research Center in Beijing. As the only developing country of the six member states (the United States, the United Kingdom, France, Germany, Japan, and China) participating in the HGP, China has undertaken 1% of the sequencing task. Professor Yang Huanming, Director of the Beijing Institute of Genomics of the CAS and Director of Beijing Huada Gene Research Center, has been engaged in genome science research. He led the completion of the sequencing task of the Chinese part of the HGP, making China a member state of this grand project known as the "moon landing of life sciences" and receiving high praise from state leaders and the international scientific community.

On July 3, 2000, the Human Genome Center of the Institute of Genetics of the CAS, which undertook the task of 1% human genome sequencing, held an award ceremony in Beijing to commend scientists and relevant cooperative units involved in the HGP. The illustration shows Li Jiayang, director of the Institute of Genetics of the CAS (Left), presenting the award flag to Yang Huanming, secretary-general of the Major Human Genome Project of the NSFC and director of the Human Genome Center of the Institute of Genetics of the CAS.

## The Cloned Cattle Kangkang and Shuangshuang

In November 2001, the first case of somatic cloned cattle Kangkang and the second case, Shuangshuang in China, led by Dong Yajuan and Bai Xuejin, were born at Laiyang Agricultural College successively. Kangkang was a black beef cow. It weighed 30 kg, was 64 cm long, and showed a strong heartbeat and healthy respiration at birth. Its birth was beyond the expectation of the experts present, and the delivery was smooth. Kangkang could move around well just over an hour after it was born. A few days later, Cuihua, another cow, who became pregnant at the same time as Kangkang's mother Guihua, also delivered a female calf called Shuangshuang. Kangkang and Shuangshuang were the first cloned cattle cultivated by using cloning technology independently in China, which greatly promoted the development of animal cloning technology in China. They were also the only two healthy somatic-cloned cattle in China back then and also the world's only twin-cloned cattle, which means that the success rate of cattle cloning in China has reached the world's advanced level.

Zhao Zhongxian (1941–), a famous superconducting expert member (academician) of the Academic Divisions of the CAS. In the 1980s, the research on superconducting high tech in China attracted wide attention. The illustration shows Zhao Zhongxian (left) testing the laboratory's electromagnetic properties of superconductor samples.

## Zhao Zhongxian Bore Fruit in High-Temperature Superconductivity Research

Superconductivity is one of the most wonderful phenomena in the physical world. Typically, when electrons move in metals, they bounce and lose energy due to the imperfections of the metal lattice (such as defects or impurities); that is, there is resistance. In the superconducting state, electrons can move forward without fetters because electrons become pairs when the temperature is lower than a specific temperature. At this time, if the metal wants to impede its movement, it needs to break up the electron pairs first. When the temperature is lower than a certain figure, more energy will be required to do the trick so that the electron pairs can move smoothly. On March 28, 1986, a scientific research group led by Zhao Zhongxian from the Institute of Physics of the CAS reported that the critical temperature of high-temperature superconductivity of fluorine-doped praseodymium oxide iron arsenic compounds could reach 52 K (−221.15°C). On April 13, the research team made a new discovery that the superconducting critical temperature of fluorine-doped praseodymium oxide iron arsenic compounds could be further increased to 55 K (-218.15°C) if they acted under pressure. In addition, the scientific research team led by Wen Haihu from the Institute of Physics of the CAS also reported that the superconducting critical temperature of strontium-doped lanthanum oxide iron arsenic compounds is 25 K (-248.15°C).

## (4) Basic Research of Applied Science and Technology

Basic research is the source of development for science, technology, and the economy and the precursor of new technologies and inventions. During the period of the *Fifteen-Year Science and Technology Plan*, it was required to, revolving around the national strategic needs and international scientific frontiers, concentrate on supporting research on major scientific issues in the national economy, social development, and national security, strengthen applied basic research, and strive to make new progress in genomics, information science, nanoscience, ecological science, earth science, and space science; steadily promote discipline construction, and strengthen frontier and cross research and accumulation in critical fields of basic disciplines such as mathematics, physics, chemistry, and astronomy; create an environment of free thinking, truth seeking and continuous progress, and encourage scientists to conduct exploratory research.

The goal was to continue to cultivate a top-notch talent team, enhance China's innovation capability in basic research, and strive to climb the world's scientific peak so that within 10–15 years, China become one of the world's middle scientific powers, and able to independently solve significant scientific and technological problems in economic, social development and national security.

### Achievements in the Study of Weng'an Animal Fossils

In 1998, Chen Junyuan et al. discovered fossil types such as cyanobacteria, multicellular algae, acritarch, resting eggs and embryos of metazoans, suspicious sponges, tubular metazoans, and tiny bilaterally symmetric metazoans in the phosphorite in Doushantuo Fm (about 600 million years ago) of the Sinian System. This fauna is considered the best-preserved evidence of multicellular plants with cellular structure in all fossil records. Weng'an, a mysterious land, has become a window for exploring animals' origin and early evolution from different perspectives of molecules, cells, ontogeny, and adult morphology.

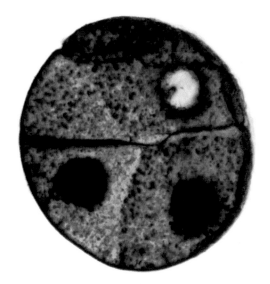

Weng'an animal fossils.

## First Fine Map of Rice Genome

China announced the implementation of the super hybrid rice genome project in May 2000. On December 12, 2002, the CAS, the Ministry of Science and Technology of the PRC, the State Development Planning Commission, and the NSFC jointly held a press conference to announce the completion of the fine map of China's rice (indica) genome. Unlike other major rice genome projects in the world, the sequencing materials of Chinese scientists were super hybrid rice provided by academician Yuan Longping. Scientists have completed their work with high efficiency and quality in a year or so. The work frame diagram of the rice genome they have drawn covers the whole rice genome and over 92% of rice genes. Their workload is equivalent to 10 repeated determinations of the rice genome. They were surprised to find that the total number of genes in the rice genome was 46,022–55,615, almost twice that of human genes.

## Qinshan Nuclear Power Plant

China proposed to build nuclear power plants in the early 1970s. In February 1970, Premier Zhou Enlai clarified that China should develop nuclear power plants. On December 15 of the same year, Premier Zhou listened to the report on the principle scheme of atomic power plants and formulated the guiding principle of "safe, applicable, economic and self-reliant" for constructing nuclear power plants. In November 1981, the State Council approved the first nuclear power plant project. In November 1982, it approved the final site selection of this project in Qinshan Mountain, Haiyan, Zhejiang Province.

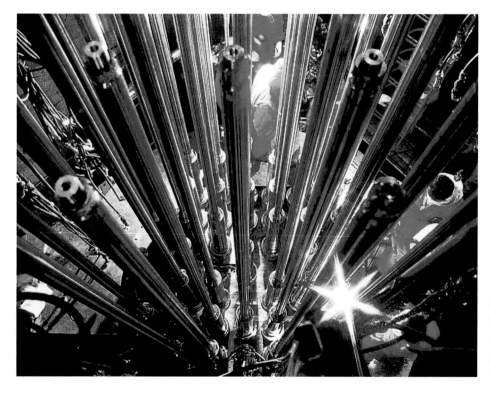

Qinshan Nuclear Power Plant is connected to the grid for power generation.

In the early morning of December 15, 1991, the Qinshan Nuclear Power Plant, the first designed and built by China, was connected to the grid for power generation, thus ending the history of zero nuclear power in mainland China. The successful construction of the Qinshan Nuclear Power Plant marks that China has mastered atomic power technology, thus becoming the seventh country in the world that can independently design and manufacture nuclear power plants after the United States, the United Kingdom, France, the Soviet Union, Canada, and Sweden. That China has overcome many difficulties and independently built the Qinshan Nuclear Power Plant is a demonstration of its comprehensive national strength and matters a great deal to solve the imbalance between energy supply and demand in China, especially in the eastern coastal areas.

## Achievements in Nanotechnology

Nanoscience and technology are some of the globally recognized frontier fields today. Nanotechnology emerged in the 1990s. It is a new technology that studies the laws and characteristics of the movement of electrons, atoms, and molecules in the space of 0.10–100 nanometers (i.e., one billionth of a meter). It uses single atoms and molecules to make matter. Its emergence will give rise to new science and technology, such as nanoelectronics, nanomaterial science, nanomechanics, etc. China's nano research level and R&D capability have gradually entered the international mainstream direction and have made outstanding achievements, ranking at the forefront of international science and technology.

The researchers from the State Key Laboratory of Rapid Solidification Nonequilibrium Alloys of the CAS are conducting experiments on the preparation of nanomaterials.

# (5) Agricultural Technology

Agricultural technology mainly refers to the cultivation of improved varieties of crops and corresponding cultivation techniques, such as transgenic wheat resistant to barley yellow dwarf virus (BYDV), transgenic potato resistant to bacterial wilt, insect-resistant cotton, and transgenic rice that have achieved success for the first time in the world. The research of the two-line method (utilization of heterosis between subspecies) in China is in the leading position in the world.

## Transgenic Wheat Resistant to BYDV

In late November 1995, the world's first transgenic wheat resistant to BYDV was cultivated by the research team led by Cheng Zhuoming of the Institute of Plant Protection of the Chinese Academy of Agricultural Sciences and passed the expert appraisal in Beijing. After three years of endeavor, the research team measured the nucleotide sequence of the yellow dwarf virus coat protein gene, deciphered its genetic code, and performed artificial synthesis. Subsequently, the research team introduced the artificially synthesized virus coat protein gene into common wheat, employing a pollen tube pathway and particle bombardment. Three detection methods proved that the foreign gene existed in the transgenic wheat and was stably inherited by the third generation. China has obtained virus-resistant transgenic wheat for the first time in the world, laying a solid foundation for breeding wheat with disease resistance.

Transgenic wheat resistant to BYDV.

Hybrid rice cultivated through a two-line method.

## The Research of Two-Line Hybrid Rice in China Leads the World

In 1998, Xue Guangxing, a researcher at the Institute of Crops of the Chinese Academy of Agricultural Sciences, and other researchers, after 11 years of studies, uncovered the mystery that hindered the promotion of two-line hybrid rice in China. They discovered that it is more important to study the succession and variation of light and temperature sensitivity and that the non-sensitivity of sterile line plants to light and temperature causes a high risk of two-line hybrid rice seed production and reproduction. For years, the popularization of two-line hybrid rice has been hindered because of its low purity and unstable seed-setting rate. In this regard, the Crop Institute of the Chinese Academy of Agricultural Sciences has conducted in-depth research, where researchers selected a Co3 strain from the core seed Peiai 64S currently used in production. Its heading date in Beijing is one week later than Peiai 64S, but it exhibited better purity and sterility stability than Peiai 64S. Xue believes that the significance of this study lies in that it can improve the quality of photo-sensitive genic male sterile lines, further enhance the purity and breeding success rate of sterile male lines and hybrid rice, speed up seed breeding, and reduce the risk and cost of two-line hybrid rice production.

# (6) Medical Domain

Hepatitis B vaccine, interferon, interleukin-2, and basic fibroblast growth factor have been industrialized, and many Chinese indicators have reached or exceeded the international indicators of similar products. In terms of biotechnology, many are among the best in the world.

## Interferon, the First Genetic Engineering Drug in China

Malignant tumors and hepatitis B, among other diseases, seriously endanger people's health. Using genetic engineering to produce drugs and vaccines to prevent and treat difficult diseases is a hotspot of modern biotechnology. The genetic engineering interferon α 1b, developed in China, is the first genetic engineering drug in the world that uses the cloning and expression of interferon genes from healthy Chinese people's white blood cells. α- Interferon is currently recognized as the most effective anti-hepatitis virus drug in the world. It is also the first industrialized product in the biotechnology field of Project 863 in China and the first genetic engineering drug approved by the Ministry of Health in China. It has been included in the first national Torch Program and listed as China's first class-I new drug. Interferon is an endogenous drug that can treat many difficult diseases. And endogenous drugs are a new direction of medical development.

In experimental medicine, China has researched hormone metabolism, drug metabolism, immune mechanism, and traditional Chinese medicine, providing valuable information. The illustration shows the radiation source for studying and preparing radioactive instruments.

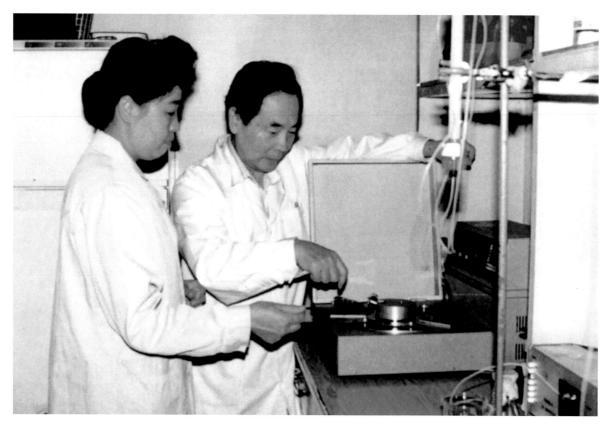

Hou Yunde (1929–), an expert in medical virology, is an academician of the CAE. Hou (right) successfully developed the genetic engineering drug interferon.

# (7) Resource Exploration

To increase the oil and gas supply and ensure national energy security, we must seek new ways of oil and gas exploration, strengthen oil and gas exploration, and promote oil and gas discovery. We mainly carried out the systematic study of oil and gas resources in the Tarim Basin and the exploration study of the East China Sea gas field. The nonferrous metal exploration focused on the Sanjiang (three-river) Metallogenic Belt in Southwest China. In this area, the Nu River, Lancang River, and Jinsha River (Yuan River) flow, including southern Qinghai, eastern Xizang, western Sichuan, western Yunnan, and Xinjiang. The metallogenic belt is located at the junction of the Indian plate and the Yangtze plate, with complex geological structures, diverse sedimentary formations, frequent magmatic activities, and strong metamorphism.

## Investigation of Sanjiang Ore Belt in Southwest China

The Nu River, Lancang River, and Jinsha River-Hong River basins in southwest China are commonly known as the Sanjiang Region. Geologically, it is located in the zone with the most intense compression, fold, and nappe east of the Tethys Himalayan tectonic domain. The complex geological structure in the area forms large-scale numerous arc-shaped deep fractures and great fractures. There are abundant mineral deposits of nonferrous metals, precious metals, rare metals, and nonmetals. A series of distinctive ore concentration areas are distributed, such as Yulong with mainly copper, Jinding with mainly lead and zinc, Yaniuping with mainly rare earth, Ailao Mountain with mostly gold, etc. Based on summing up the essential characteristics of these ore concentration areas, Chinese researchers discussed the basic pattern of the distribution of ore concentration areas (i.e., two horizontal lines, two vertical lines, two oblique lines, and one point) and their tectonic background. On top of that, they proposed three ore concentration areas whose prospect can be expanded and three potential ore concentration areas and discussed new prospecting directions.

The illustration shows scientific and technological personnel investigating the Sanjiang ore belt.

## Exploration of Oil and Gas Fields at Tarim Basin and the East China Sea

The Tarim Basin is China's largest oil and gas-bearing basin, with 17.8 billion tons of oil and gas resources, a very broad exploration prospect. However, the surface condition of the basin is highly adverse, and its geological structure is extremely complex, making it one of the most challenging areas for oil and gas exploration and development in the world. In 1989, China's petroleum scientific and technological personnel launched a mass campaign in the Tarim Oilfield. Their discovery of the Kela 2 gas field in 1998 facilitated the establishment and decision of the West-East Natural Gas Transmission Project, ushered in the era of large-scale utilization of natural gas in China, and made positive contributions to changing the country's

Tarim Basin oil exploration site.

energy structure. Since 1974, China has been conducting oil and gas exploration in the East China Sea and has discovered several oilfields. In 1995, the New Star company succeeded in trial drilling in the Chunxiao area. Chunxiao Oil and Gas Field lies in the East China Sea, 500 kilometers southeast of Shanghai and 350 kilometers from Ningbo. Experts call its location the "West Lake Depression Domain in the East China Sea." This offshore oil and gas field consists of four oil and gas fields, covering an area of 22,000 square kilometers.

The largest natural gas processing plant in Asia of Tarim Oilfield—Central Processing Plant in the Kela Operation Area.

Qaidam Basin stores an abundance of oil. In January 1959, Qinghai Petroleum Exploration Bureau changed its name to Qinghai Petroleum Administration Bureau and concentrated its main forces on drilling in Cold Lake. In that year, its crude oil exceeded 300,000 tons.

## (8)  Environmental Investigation

The complex coupling system composed of solid earth, atmosphere, hydrosphere, and biosphere is studied as a whole to provide basic data and a theoretical basis for solving significant problems such as national resources, energy, environment, and natural disasters. Ecology research focuses on the coevolution of systems, the mechanism of degraded ecosystems, and the formation of optimized artificial systems, thus contributing to improving the environment and promoting social development.

Yarlung Zangbo River Grand Canyon.

### Scientific Investigation of the Yarlung Zangbo River Grand Canyon

In 1994, Chinese scientists assembled a scientific expedition to investigate the Yarlung Zangbo River Grand Canyon, lifting its mysterious veil. The Yarlung Zangbo River Grand Canyon stands on the Qinghai-Xizang Plateau with an average altitude that exceeds 3,000 meters. It is steep and deep. Its downward erosion reaches 5,382 meters. With nine natural vertical zones, from alpine ice and snow zone to tropical monsoon forests in low valleys, it holds the most complete mountainous vertical natural zones in the world. It gathers an assortment of biological resources, including 2/3 of the known higher plant species on the Qinghai-Xizang Plateau, 1/2 of the known mammals, 4/5 of the known insects, and 3/5 of the known macrofungi in China, altogether the most in the world.

## Arctic Scientific Expedition

On July 1, 1999, the first batch of Chinese Arctic scientific research teams set out from Shanghai on the Snow Dragon scientific research ship, crossed the Sea of Japan, Zonggu Strait, Okhotsk Sea, and Bering Sea, entered the Arctic Circle twice, reached the Chukchi Sea, the Canadian Basin, and the multi-year sea ice area, successfully completing the planned on-site scientific investigation tasks set up within the three major scientific objectives, and obtaining heaps of extremely valuable samples, data, and materials. After 71 days, the Snow Dragon ship, loaded with the fruitful results of China's first Arctic scientific expedition, safely covered 14,180 nautical miles (1 nautical mile equals 1.852 kilometers, the same as below), sailed 1,238 hours and returned to Xinhua Wharf of Shanghai Port on September 9, 1999. On the way back, it once docked at Nome Port, Alaska, for oil and water replenishment. The main working areas of this expedition were the Bering Sea and the Chukchi Sea.

Apart from ice avalanches, extremely cold weather, and snowstorms, Arctic exploration's dangers also include the threat of polar bears. The illustration shows the scientists observing the meteorological gradient on the Arctic ice floe. The gun is loaded to protect them from polar bears.

## (9) R&D of Large Complete Sets of Equipment

Large complete sets of equipment mainly included 20 million-ton large open-pit mine complete sets of equipment, 600,000-kilowatt nuclear power units, 500,000-volt DC power transmission and transformation complete sets of equipment, heavy haul train complete sets of equipment, 300,000-ton ethylene complete sets of equipment, etc.

### 300,000-Ton Ethylene Chemical Plant

Since the reform and opening up, China's petrochemical industry has been developing rapidly. As the petrochemical industry leader, the ethylene industry's development in China has witnessed fierce competition for projects. The ethylene industry is a heavy chemical industry with high input, output,

technology content, and added value. Its layout must meet the needs of the national economy's overall development and match the regional economy's characteristics. When developing the ethylene industry, China, having absorbed the experience of the ethylene industry layout of the United States, Japan, South Korea, and other countries, designed a reasonable layout for it, and adopted an upsizing, large-scale, base-based and park-based development model, thus gradually achieving effective and rapid development.

Ningxia 300,000-ton ethylene chemical plant under construction.

National key project—Golmud Oil Refinery.

## 20-Million-Ton Large Mining Equipment

China, as one of the largest mining countries in the world, has an output of metal ores that is at the global forefront. In 1999, its ore output from metal mines exceeded 300 million tons, including 209 million tons of iron ore and 93 million tons of nonferrous metal ore. The progress of the mining industry mainly depends on the development of mining equipment. Chinese metal mines, especially underground ones, had a wide gap compared to advanced foreign mining equipment. Since the reform and opening up, through the technical breakthrough of major mining equipment and the introduction and absorption of foreign advanced technical equipment, China's mining equipment has been dramatically improved, with a certain scale and level. China has been able to manufacture complete sets of various major mining equipment for open-pit mines with an annual output of 10 million tons and underground mines with an annual output of one million tons. As China joined the World Trade Organization and the knowledge economy era in the 21st century arrived, developing mining equipment in China's metal mines has welcomed excellent opportunities.

Large mining machinery in mines.

This blue-and-white train is a quasi-high-speed train designed and manufactured by China in the 1990s.

## (10) Transportation Technology

Transportation technology mainly includes railway operation management and control technology, railway, and highway passenger transport technology, new locomotive technology, high-grade highway and road materials technology, civil aviation navigation communication, air traffic control, and operation management technology, main trunk aircraft design and manufacturing technology, inland waterway dredging equipment, new inland ship technology, port handling technology, etc.

## (11) Raw Material Technology

Raw material technology mainly includes localization of large-type chemical catalysts, coal chemical technology, oxygen coal intensive smelting technology, nonferrous metal energy saving and comprehensive utilization technology, energy saving technology of building materials industry and refractory manufacturing technology, etc.

## (12) Other Technologies

Other technologies mainly include population control and eugenic technology, new disease prevention and control technology, comprehensive pollution prevention and control technology, water and soil conservation technology, monitoring and forecasting technology for major and frequent natural disasters, etc.

# Science and Technology, the Primary Productive Forces—Implementation of the *Program* and the *Outline*

In 1978, the remarkable National Science Conference opened the great prelude to the "four modernizations" of China. As the modernization of science and technology became the key to realizing the "four modernizations," the status of intellectuals was significantly improved as a part of the working class for the first time. At this conference, Comrade Deng Xiaoping made the famous assertion that "science and technology are productive forces." In 1988, he further pointed out that "science and technology are the primary productive forces." From then on, the development of science and technology has become a strategic task and a top priority.

Human society was about to enter the 21st century, and the world was undergoing great change. China faced urgent and severe challenges due to the rapid development of new scientific and technological revolutions, increasingly fierce market competition, and constantly changing international politics. It must follow the basic line of "one central task, two basic points," implement reform and opening up in an all-round way and vigorously promote the transfer of economic construction to the track of relying on scientific and technological progress and improving the quality of workers. To this end, in 1992, the State Council issued the *National Program for Medium and Long-Term Science and Technology Development* (the *Program*), and the Ministry of Science and Technology formulated the *Outline for Medium and Long-Term Science and Technology Development: 1990–2000–2020* (the *Outline*). Aiming at the prospects of medium and long-term science and technology development, both have made a macro and general statement around the critical issues of science and technology, economic and social development.

## Develop High and New Technology and Its Industry

Under the guidance of Deng Xiaoping's idea that "science and technology are the primary productive forces" and the policy of "develop high technology and industrialize" through the promotion of reform and innovation and the guidance of market mechanisms based on giving full play to the advantages and potential of China's scientific and technological forces, the commercialization, industrialization, and internationalization of high and new technology have been greatly accelerated, and a path of China's high-tech industrialization has been explored.

## Steadily Strengthen Basic Research

The development of basic research is crucial for China to achieve the goal of stepping into the ranks of scientific and technological power. Compared with the major developed countries and some newly industrialized countries, China lagged far behind in basic research. To solve this problem, it is necessary to analyze the demand for basic research in China and formulate feasible development goals.

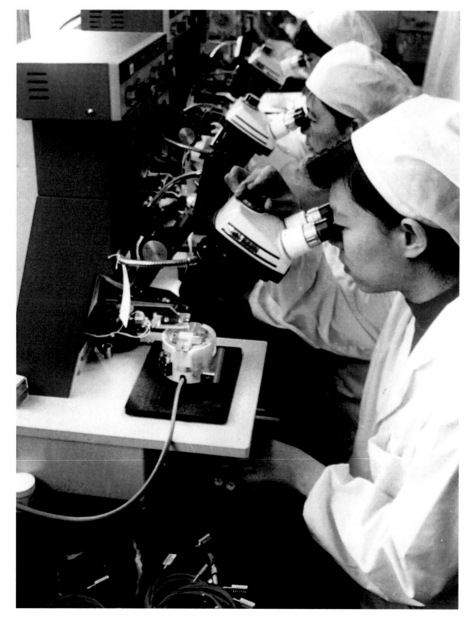

The production line in product development base 863 can produce various new and high-performance semiconductor optoelectronic devices according to market demand.

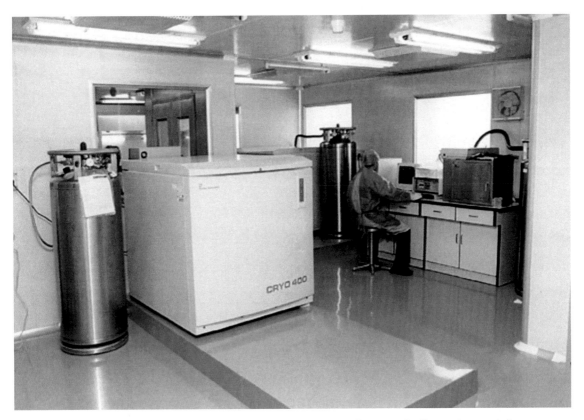

Tianjin Xiehe Stem Cell Co., Ltd. has built a production base for genetically engineered drugs.

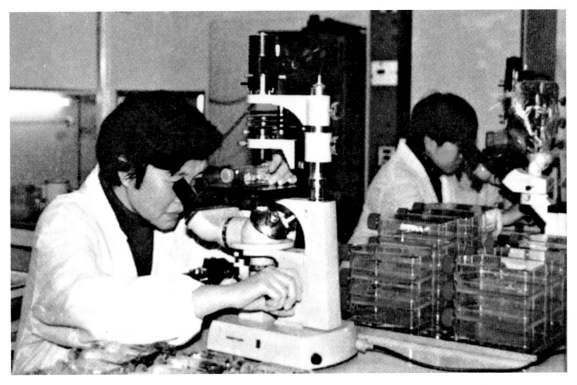

Researchers from Fudan University prepared genetically engineered cells for gene therapy.

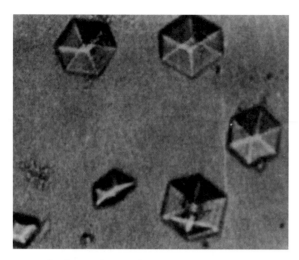

Genetically engineered insulin crystals.

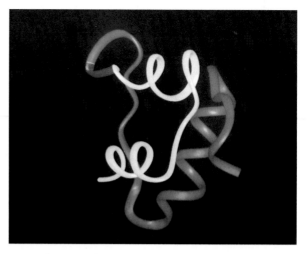

Insulin precursor prepared by genetic engineering.

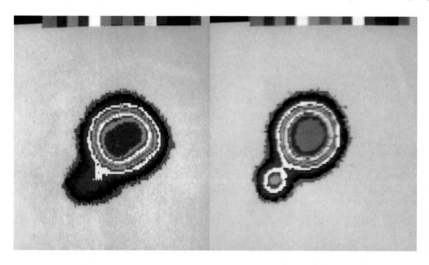

The binary star photo taken by the infrared adaptive optical imaging system of the astronomical telescope.

Infrared adaptive optical imaging system installed on 2.16 m astronomical telescope.

Controlled nuclear fusion research device—the main engine of China HL 1.

# Development of Information Science and Technology

The world has entered the information age. In the current times of rapid economic growth and fierce competition, information technology has created enormous material wealth on the one hand. In contrast, on the other, it has also triggered changes in the way of social production, lifestyle, and even thinking patterns. Faced with such a reality, China has determined intelligent technology, optoelectronic technology, information acquisition, processing technology, and modern communication technology as its research topics based on its national conditions, aiming to lay a solid foundation for the revitalization of its national information industry. It mainly includes 3-micron production process, 1-micron, and submicron process technology in microelectronics technology; R&D of an application-specific integrated circuit and key application-specific equipment; fifth-order optical communication system technology, remote sensing technology, large-scale computer system, software engineering technology, etc.

The optical fiber transmission system with a 12-channel wavelength division multiplexing+ optical fiber amplifier developed by Peking University adopts the self-developed wavelength automatic control system.

The green laser used for high-power semiconductor laser gas pump has the advantages of high output power, high efficiency, long life, small size, convenient use, etc. It is widely used in color television, laser printer, compact disc technology, medical treatment, underwater communication, submarine exploration, spectral technology, and navigation.

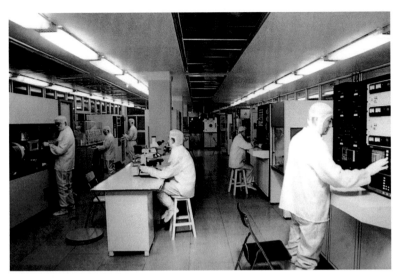

Photoelectronic technique is a cross-discipline formed by the combination of electronic technology, optical technology, and laser technology. It has a series of characteristics, such as large information communication capacity, long relay distance, large information storage density, fast information processing speed, easy-to-realize parallel and interconnected processing, high information acquisition sensitivity, anti-electromagnetic interference, and anti-radiation.

## Life Science and Biotechnology

Life science and biotechnology are widely used in China's agricultural, industrial, and security fields. Its research scope is increasingly broad, including genetic engineering, protein engineering, cell engineering, tissue engineering, and animal cloning, among many other aspects. It involves medicine, pharmacy, and marine science, among other disciplines, such as personalized disease diagnosis, treatment, prevention, etc. After a period of development, China's innovation capability in life science and biotechnology has been dramatically improved. For example, using gene transfer and integration, we have cultivated new strains of animals and plants with unique properties, making a miracle with agricultural production technology; we have bred transgenic cotton plants resistant to cotton bollworm and tested the wheat resistant to yellow dwarf disease, gibberellic disease and powdery mildew in the field; we have achieved success in fast-growing transgenic fish and test tube cattle, etc.

The fast-growing transgenic carp developed by Heilongjiang Fisheries Science Research Institute can increase production by more than 15%.

Transgenic plants cultivated in a greenhouse.

Insect-resistant transgenic cotton with gene B.T.

The insect-resistant gene was introduced into cotton to create insect-resistant cotton, which exhibits effective cotton bollworm resistance. The left of the illustration shows insect-resistant transgenic cotton. The insect's body shrinks, and the cotton wound heals.

Transgenic operation on plants can be carried out by using a gene gun.

## R&D of Energy Science

Energy is the lifeblood of the national economy. Since the reform and opening up, China's energy industry has developed rapidly, and its energy structure has also undergone great changes. For the long-term development of the country, in addition to strengthening the development of stable and high production technology of oil and gas fields, comprehensive coal mining and safety production technology, clean coal combustion technology, new dam types, and dam building technology for hydropower use, Project 863 has selected to tackle critical problems in coal magnetic fluid power generation technology and advanced nuclear reactor technology. Through the efforts of many scientific researchers, high-temperature gas-cooled reactors and fast neutron reactors have made significant breakthroughs.

Fast neutron reactor has dramatically improved the utilization rate of nuclear fuel. It can also generate nuclear fuel while generating electricity, which is of great significance for the full use of nuclear resources.

## Vigorously Performed International Cooperation in Science and Technology

In the middle and late 20th century, the development of the international situation presented two remarkable characteristics: the rapid advancement of science and technology and the globalization of the economy. Under such circumstances, there is frequent international cooperation, of which scientific and technological cooperation is essential. China's scientific and technological development process is involved in international competition. Its national scientific research institutions, with their strong comprehensive research strength, are responsible for researching a great number of major and critical scientific and technological projects and promoting the development of science and technology in China and the world through scientific and technological exchanges between the government and the people.

Shanghai Volkswagen, a Sino-German joint venture, is one of China's automotive industry's pillar enterprises. From December 1984 to the early 1990s, it manufactured nearly 400,000 cars.

China and the United States cooperated to launch satellites. The illustration shows technicians from both parties taking a group photo before the launch.

With the continuous exploration and development of international cooperation, some Chinese companies are increasingly expanding their influence overseas, and foreign business people looking for partners are pouring in.

Li Siguang's research on China's Quaternary glaciers has attracted wide attention from the geological community in China and the world. This was an academic exchange between foreign guests and Chinese scholars.

# The Strategic Decision on Invigorating China through Science and Education

IV

From May 26 to 30, 1995, the National Conference on Science and Technology was held in Beijing. The illustration shows the venue of its opening ceremony.

In May 1995, the Central Committee of the CPC and the State Council issued the *Decision on Accelerating the Advancement of Science and Technology* (referred to as the *Decision*), proposing the great strategy of "invigorating China through science and education." The Decision points out that invigorating China through science and education means comprehensively executing the idea that "science and technology are the primary productive forces," adhering to education is the foundation, placing science and technology and education in an important position in economic and social development, enhancing China's scientific and technological strength and its ability to transform it into productive forces, improving the scientific and cultural quality of the entire nation, and shifting economic construction to the track of relying on scientific and technological advancement and improving the quality of workers, to ultimately accelerating the realization of national prosperity and strength.

In 1997, the 15th National Congress of the CPC was held, further establishing the development strategies of "invigorating China through science and education" and "sustainable development." It proposed that the development of the national economy should achieve two fundamental changes; namely, the mode of economic growth should be changed from extensive to intensive, and the construction focus of the national economy should be changed to scientific and technological advancement and improving the quality of workers. It clearly put accelerating scientific and technological progress at the critical

# Full Implementation of the Strategy of "Invigorating China through Science and Education"— Implementation of *Scientific and Technological Development Plan during the Ninth Five-Year Plan Period and the Long-Term Program until 2010*

In May 1995, Comrade Jiang Zemin proposed the strategy of "invigorating China through science and education" in his speech at the National Conference on Science and Technology and established the guiding principle that science and technology and education are the means and basis of invigorating China. This guiding principle has effectively raised the awareness of cadres at all levels of the importance of science, technology, and education and deepened their understanding that science and technology are the primary productive forces. To implement the strategy of "invigorating China through science and education," we should not only give full play to the role of science, technology, and education in rejuvenating the country but also strive to cultivate the foundation of science, technology, and education.

At present, we should pay more attention to strengthening and supporting science, technology, and education to lay a good foundation for the country's recent development and long-term stable development, improve the contribution rate of producers to economic growth, and found high-tech enterprises as soon as possible; at the same time, we should strengthen the improvement of the populace's cultivation, strengthen basic education, pay attention to the training of talents, and attach importance to creative scientific research. Science, technology, and education have dual functions, providing various means for the current economic and social development and laying the foundation for sustainable and long-term development. Today's science, technology, and education can provide knowledge, techniques, and talents for the development of the economy and society, thus generating benefits. It is also a return on the previous investment in science, technology, and education.

# (1)  Science Plan Issued and Related Work

## Science and Technology Plan for Social Development

The Science and Technology Plan for Social Development aims to solve the scientific and technological problems in the fields of social development, such as environmental protection, rational exploitation and utilization of resources, disaster reduction and prevention, population control, and people's health, contribute to improving the ecological environment and the quality of people's life and health, and promote the sustainable and coordinated development of economy and society. It is a horizontal coordination plan in the matrix management system of the science and technology program.

## National Technological Innovation Plan

The National Technological Innovation Plan is one of the main plans of the National Science and Technology Program. It is a national project to guide and absorb enterprises and social forces (including talents and funds) and enhance enterprises' technological innovation capability with government funding and financial support. The plan includes technology development, high-tech industrialization, technology center construction, etc.

## National Key Basic Research Development Plan (Project 973)

The national key basic research development plan (project 973) was officially initiated in 1998. Aiming at the forefront of science and major scientific problems, it is a national plan for innovative basic research based on existing basic research work revolving around key fields, to provide substantial scientific support for solving major problems in China's economic and social development in the early 21st century, to cultivate a group of talents with extensive scientific knowledge and innovative capability, to strengthen base construction, to improve the scientific level, and to drive the comprehensive development of basic research and even science and technology in China.

## Technical Innovation Fund for Technology-Based SMEs

The technical innovation fund for technology-based SMEs is a non-profit guiding fund set up under the approval of the State Council. It supports the technological achievements that align with the national industrial technology policy, has a high level of innovation and strong market competitiveness, generates excellent potential economic and social benefits, and is expected to become emerging industries. It adheres to the principles of scientific evaluation, favors the best candidates, fairness, transparency, and dedicated use, introduces a competition mechanism, and implements the project evaluation and bidding system of innovation funds.

## Knowledge Innovation Project

The knowledge innovation project is an integral part of the construction of the Chinese innovation system, marking that the reform of China's science and technology system has entered a new stage and a new development period. With high goals, starting points, and requirements, the CAS's first batch of knowledge innovation pilots has been conducted step by step in accordance with the idea of unified planning, step-by-step implementation, key breakthroughs, and comprehensive promotion.

## Basic Science and Technology Projects of Central Scientific Research Institutes

In 1999, the central scientific research institutes' basic science and technology project commenced. With the central scientific research institutes as the main body, it has driven the construction of basic science and technology work bases, promoted the improvement and development of the basic science and technology work system, gradually established and improved the sharing mechanism of resources and achievements, and ensured the realization of social sharing.

## Special Funds for Technology R&D of Scientific Research Institutes

In 1999, some funds were pooled from the reduced scientific expenditure to set up a special fund for technology R&D of scientific research institutes. It was dedicated to supporting the application R&D aimed at developing high-tech products or engineering technologies by the central technology development research institutions (including the original ones transformed after 1999).

## Action Plan for Invigorating Trade through Science and Technology

The action plan for invigorating trade through science and technology was proposed in 2002 by the Ministry of Foreign Trade and Economic Cooperation, the Ministry of Science and Technology, the State Economic and Trade Commission, the Ministry of Information Industry, the Ministry of Finance, the General Administration of Customs, the Tax Bureau and the Quality Supervision and Inspection Bureau. It aims to improve the policy environment of high-tech import and export mainly by giving play to the guidance and service functions of government policies; accelerating the structure adjustment of import and export commodities, and promoting the internationalization of China's high-tech industries; promoting the export of high-tech products with independent intellectual property rights, and increase the proportion of technology-intensive products in export products; based on exerting comparative advantages, create new competitive advantages, and ultimate realize the strategic transformation of China's foreign trade development model.

Accelerate the learning and absorption of the imported technology, and localize the license model drawings and process documents strictly following Mercedes-Benz standards. This is an elaborate assembly of the North Benz truck by technicians.

## National University Science Park

The National University Science and Technology Park is an institution that, relying on research universities or university groups, combines the comprehensive technological advantages of universities, such as talents, technology, information, experimental equipment, books, and materials, with technological advantages to serve technological innovation and achievement transformation. In September 1999, the Ministry of Science and Technology and the Ministry of Education jointly issued the *Notice on Organizing the Pilot Construction of University Science and Technology Parks*. Since then, the pilot work of national university science and technology parks has officially kicked off, and its construction has been jointly promoted at the national level.

## The Social Welfare Special Project of the Scientific Research Institute

The social welfare research project of scientific research institutes was organized and implemented by the Ministry of Science and Technology in 2000. It focused on supporting the construction of several social welfare research bases, forming a social welfare research network, providing technical support for socially sustainable development and public welfare services, and promoting the improvement of sustainable innovation capability and level of social welfare research.

## Three Gorges Resettlement Science and Technology Development Project

The Three Gorges Resettlement Science and Technology Development Project, launched in 1996, was organized and implemented by the State Scientific and Technological Commission and the Resettlement Development Bureau of the Three Gorges Project Resettlement Committee of the State Council (referred to as the Three Gorges Resettlement Bureau of the State Council). Its task was to figure out the common and key technologies in the economic development and ecological construction of the Three Gorges area through the development, promotion, and introduction of advanced practical technologies, to promote the development of regional pillar industries in the reservoir area, to cultivate characteristic emerging industries, to restore and govern the ecological environment in the reservoir area, and to promote its informatization and modernization.

"One road in front of the house, gardens behind it." The two banks of the Yangtze River are where the world-famous Three Gorges Project is. And the development planning of the immigrant villages on both banks is improving. The illustration is the Immigrant New Village of Yaoping, Yunyang County, which has already begun to take shape.

## Special Operation for the Development of the Western Region in China

To coordinate with the overall plan for the development of the western region in China, the Ministry of Science and Technology organized and conducted the special operation. It was carried out step by step with scientific planning as the top priority and key points highlighted; it focused on ecological environment construction, and it aimed to strengthen technology integration, demonstration, and promotion, thus promoting work in the western region by drawing upon the experience gained on key points; great importance was attached to the combination of resources advantages and scientific and technological advantages in the western region, to actively promote cooperation, and strengthen the construction of scientific and technological capabilities and innovation environment. This operation was mainly organized and implemented through the national key science and technology research plan, the basic research major project plan, and Project 863, among other national major science and technology programs.

## China Association for Science and Technology (CAST) Set Up the Academic Annual Conference System

In 1999, the fourth plenary meeting of the fifth session of the CAST decided to establish its annual academic conference system, which aimed to implement the strategy of invigorating China through science and education and the strategy of sustainable development and organize high-level, open, and interdisciplinary academic exchanges in the process of establishing and improving the Chinese innovation system. In 2006, the annual academic conference of the CAST was renamed as the annual conference of the CAST, establishing the orientation of the conference as "popularizing science, connecting the disciplinaries, and serving the hosting venue" and achieving a total transformation. The conference is held once a year to serve the public, scientific and technological workers, governments, and enterprises. It strives to build three platforms of academic exchange, science popularization, and decision-making consultation, thus bridging the communication and interaction between scientists and the public, scientists and governments, and scientists recognized within the science community and all sectors of society.

List of the host cities of the annual conference of the CAST and its themes.

| 1999 | 1st | Hangzhou, Zhejiang Province: Scientific and Technological Advancement and Economic and Social Development Oriented at the 21st Century |
| 2000 | 2nd | Xi'an, Shaanxi Province: Development of the Western Region in China: Science and Education as the Top Priority and Sustainable Development |
| 2001 | 3rd | Changchun, Jilin Province: New Century, New Opportunities, New Challenges—Knowledge Innovation and Development of the High-Tech Industry |
| 2002 | 4th | Chengdu, Sichuan Province: Join the WTO, Chinese Science and Technology and Sustainable Development—Challenges and Opportunities, Responsibilities and Countermeasures |
| 2003 | 5th | Shenyang, Liaoning Province: Building a Well-Off Society in an All-Round Way: The Historical Responsibility of Chinese Scientific and Technological Workers |

| 2004 | 6th | Bo'ao, Hainan Province: People-oriented and Coordinated Development |
| 2005 | 7th | Urumqi, Xinjiang Uygur Autonomous Region: The Concept of Scientific Development and Sustainable Utilization of Resource |
| 2006 | 8th | Beijing: Improve the Science Literacy of the General Public and Make China Innovative |
| 2007 | 9th | Wuhan, Hubei Province: Save Energy, Protect the Environment, and Develop Harmoniously |
| 2008 | 10th | Zhengzhou, Henan Province: Scientific Development and Social Responsibilities |
| 2009 | 11th | Chongqing: Independent Innovation and Continuous Growth |
| 2010 | 12th | Fuzhou, Fujian Province: Change of Economic Development Mode and Independent Innovation |
| 2011 | 13th | Tianjin: Scientific and Technological Innovation and Development of the Strategic Emerging Industries |
| 2012 | 14th | Shijiazhuang, Hebei Province: Scientific and Technological Innovation and Adjustment of the Economic Structure |
| 2013 | 15th | Guiyang, Guizhou Province: Innovation-driven and Transformative Development |
| 2014 | 16th | Kunming, Yunnan Province: Openness, Innovation, and Industrial Upgrade |
| 2015 | 17th | Guangzhou, Guangdong Province: Innovation Goes First |
| 2016 | 18th | Hangzhou, Zhejiang Province: Reform and Opening Up, Innovation-led |
| 2017 | 19th | Changchun, Jilin Province: Innovation-Driven and All-Round Rejuvenation |
| 2018 | 20th | Kunming, Yunnan Province: Openness, Innovation, and Industrial Upgrade |
| 2019 | 21st | Harbin, Heilongjiang Province: Reform and Opening Up, Innovation-Led—Science and Technology Propels the All-Round Rejuvenation of the Northeast in the New Times |

## (2) Scientific and Technological Advancement Increases Agricultural Production

Under the guidance of the strategy of "invigorating China through science and education," during the Ninth Five-Year Plan period, China has achieved fruitful results in promoting agricultural construction through scientific and technological advancement, and its agricultural scientific and technological strength has dramatically improved. Seven hundred sixty agricultural scientific and technological achievements passed the ministerial appraisal, winning 248 national science and technology advancement awards and 42 national technology invention awards.

### Strengthen the Combination of Biotechnology and Conventional Techniques

China has cultivated a significant number of new crop varieties with high quality, high yield, and multi-resistance and has sifted out a group of fine germplasm resources, thus improving the overall breeding level. Statistics show that during the Ninth Five-Year Plan period, China has bred 18 major crops of 411 varieties, including rice, wheat, corn, and cotton, 276 subsidized varieties, and 719 superior breeding materials. Major breakthroughs were made in the research of super rice, whose yield in the experimental field was nearly 800 kg per mu; its research and application of univalent and bivalent transgenic insect-resistant cotton reached the advanced international level; its research on the large-scale high-yield supporting technology system of rice, wheat, corn, soybean, and cotton has been comprehensively developed, providing technical support and reserves for future domestic food security.

### A Major Breakthrough in the Energy-Saving Technology of Solar Greenhouse

"Factory farming," which focuses on protected horticulture, integrates energy-saving cultivation supporting technologies in solar greenhouses from the aspects of improving structure and performance, screening excellent varieties, increasing organic fertilizer and carbon dioxide, and integrated pest control, and promotes the production of protected vegetables, fruit trees, and flowers in winter. By the end of the Ninth Five Year Plan period, the area of horticultural facilities in China had reached 22.5 million mu, and the per capita share of protected vegetables accounted for 20% of the total per capita.

### Fruitful Results in Large-Scale Breeding of Main Livestock and Poultry and the Diagnosis and Monitoring Methods of Their Diseases

During the Ninth Five-Year Plan period, China has obtained 19 patents, new veterinary drugs, and new breeds (strains) of livestock and poultry for large-scale breeding of main livestock and poultry. Among them, the high-yield supporting system of the Xinyang brown layer package has passed the national examination and approval and has bred Chinese Simmental; diagnostic methods for African swine fever PCR and ELIA were established for the first time, and the production capacity of test kits was formed; we have completed the isolation and identification of avian influenza epidemic strains in China, the development and application of avian influenza recombinant nucleoprotein diagnostic antigen, and established avian influenza immunoenzyme diagnostic methods and techniques.

Wang Tao (right), an academician of the CAE and researcher of the Chinese Academy of Forestry Sciences, has been engaged in the research of asexual propagation of forest trees for a long time, invented the three-dimensional cultivation technique of plants, and committed to the research and promotion of ABT rooting powder series technology.

## ABT Plant Growth Regulator

ABT rooting powder series is a new type of broad-spectrum and efficient plant growth regulator developed by academician Wang Tao of the Chinese Academy of Forestry Sciences. It breaks through the traditional way of providing exogenous hormones for plant growth and development from the external world. By strengthening and regulating the content of endogenous plant hormones and the activity of essential enzymes and promoting the synthesis of biological macromolecules, it can induce the formation of plant adventitious roots or buds, regulate the intensity of plant metabolism to improve the survival rate of seedling raising and afforestation and the crop yield, quality, and resistance. The ABT rooting powder series has been included in the national key promotion program for scientific and technological achievements since 1989, and its application scope has continued to expand. From seedling cuttage, seeding planting, seedling transplanting, afforestation, and air seeding and afforestation to crops, vegetables, fruit trees, flowers, medicines, and other unique economic plants, it has generally improved the survival rate of seedlings and produced a significant effect of increasing production. "Promotion of ABT Rooting Powder Series" won the Special Prize National Prize for Progress in Science and Technology in 1996. In 1997, it won the Ho Leung Ho Lee Science and Technology Progress Award, and subsequently, it won 28 major domestic and international awards in total.

## (3) Scientific and Technological Advancement Promotes Industrial Development

Under the guidance of the strategy of "invigorating China through science and education," during the Ninth Five-Year Plan period, China has achieved fruitful results in promoting industrial development through scientific and technological advancement:

- The industrialization of digital program-controlled exchanges, oxygen coal-enhanced ironmaking technology, nickel-hydrogen batteries, amorphous materials, and so on has made significant achievements.
- Through scientific and technological innovation, the Three Gorges Project, integrated circuit, Qinshan Nuclear Power Plant Phase II, and other projects have conquered key technologies and mastered the design and manufacturing technology of complete sets of technical equipment.
- Computer-aided design (CAD), computer-integrated manufacturing systems (CIMS), and other major generic technologies have greatly improved the innovation capability of enterprises.

### Rapid Development of Aerospace Technology

On December 21, 2000, the second Beidou navigation test satellite developed by China was successfully launched. Together with the first Beidou navigation test satellite launched on October 31, 2000, they are the Beidou navigation system. This marks China's first generation independently developed satellite navigation and positioning system. After completion, it mainly renders navigation services for highway, railway, and maritime operations, among other fields, and plays a positive role in China's domestic economic construction.

On January 10, 2001, China's self-developed Shenzhou 2 took off at Jiuquan Satellite Launch Center and successfully entered its scheduled orbit. On January 16, this unmanned spacecraft returned accurately and safely landed. This is the first launch of the Chinese space programs in the 21st century and the second flight test of its manned space program. It marks a significant step forward for China to achieve manned flight.

At 0:34 on May 25, 2003 (Beijing time), China successfully launched the third Beidou I navigation and positioning satellite into space with the Long March 3A carrier rocket at Xichang Satellite Launch Center. The illustration shows the command and control hall of the Xichang Satellite Launch Center.

The Shenzhou Spacecraft was launched at 6:00 a.m. on November 20, 1999, from Jiuquan Satellite Launch Center. The launch mission was undertaken by the Long March 2F manned space rocket, which was improved based on the Long March 2 strap-on rocket.

At one o'clock sharp on January 10, 2001, China's self-developed Shenzhou 2 unmanned spacecraft took off from the manned space launch site of Jiuquan Satellite Launch Center. The picture shows the Shenzhou 2 spacecraft and the Long March 2F carrier rocket, having docked, were transported vertically to the launch station via the mobile launch platform.

## Oxygen Coal Intensified Ironmaking Technology

The national key scientific and technological project of the Eighth Five-Year Plan, the new process of oxygen coal intensified ironmaking in a blast furnace, undertaken by 15 units, including Anshan Iron and Steel Group Corporation (abbreviated as AnSteel), has been completed on schedule after five years of endeavor, reaching the expected objective. From August 21 to November 20, 1995, the industrial test of oxygen coal intensified ironmaking with a coal ratio of 200 kg/t was conducted in AnSteel No. 3 blast furnace. The trial achieved the expected goal: an average coal ratio of 203 kg/t, an oxygen enrichment rate of 3.42%, and a coal-to-coke replacement ratio of 0.8, realizing intensive

Oxygen coal intensified ironmaking technology.

smelting of a high coal ratio under a low oxygen enrichment rate. Through technical development and tackling key problems, a breakthrough has been made in the four aspects of the coal injection volume index, coal injection safety technology, coal injection process, and equipment level, and coal injection-related technology, and the research on the application theory of oxygen coal intensified ironmaking has also made new progress. Also, the complete set of blast furnace coal injection technology in China has taken a great leap, and the new process of blast furnace oxygen coal intensified ironmaking has reached the world's advanced level.

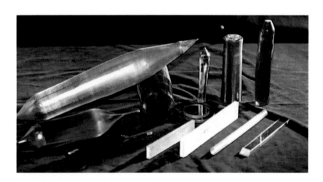

Various YAG crystals developed by the 11th Research Institute of China Electronic Technology Group Corporation.

## Breakthrough in Amorphous Materials Research

Materials are an important symbol of the progress of human civilization. And new materials are the basis and precursor for the development of high-tech. Their research, development, and application level reflects a country's overall scientific, technological and industrial levels. Therefore, every country has put its research on new materials in a prominent position. Amorphous material, also known as non-crystalline material, is a kind of artificial material produced by quickly cooling and solidifying conventional molten metal. The new amorphous and nanocrystalline materials with unique magnetic induction properties developed in China have broken through the critical technology of continuous and stable production of 22-micron ultra-thin ribbon materials and built a production line with an annual output of 500 tons of ultra-thin ribbons; amorphous filament materials with high sensitivity to force and magnetism have been developed. These new materials can be widely applied in household appliances, communication products, and automotive safety.

The largest automatic single-crystal furnace developed by the 11th Research Institute of China Electronic Technology Group Corporation produces high-quality sizable artificial crystals.

The laser pedestal method crystal fiber growth furnace developed by Tsinghua University has reached the advanced international level and produced a variety of single-crystal optical fibers.

## Tackling Technical Challenges

Revolving around the construction of the Three Gorges Hydropower Station, China has conducted studies on key equipment of the Three Gorges Project, high dam engineering technology, roller-compacted concrete high dam construction technology, and ecological reconstruction technology of the Three Gorges area. The key technology and test projects of 500 kV compact transmission lines have laid a solid technical foundation for the construction of the Three Gorges Power Transmission Project and the West-East Power Transmission Project; the 400 MW evaporative cooling water-turbine generator set has been successfully developed and applied to the construction of Lijiaxia Hydropower Station, securing China a leading position in this technology in the world; the implementation of projects such as Development of Key Technologies for High-Speed Railway, Preliminary Study on High-Speed Railway Engineering Construction and 200 km/h Electric Train Set has provided technology and equipment for speed increase and safe high-speed operation of the entire railway lines; major breakthroughs have been made in the development of ethylene-making technology via large-scale steam cracking, giving birth to the 60,000 ton/year ethylene cracking furnace and related technologies with independent intellectual property rights and international advanced level, thus providing technical support for the huge ethylene project construction of the Tenth Five-Year Plan in China.

## Significant Benefits in National CAD/CIMS Application Project

CAD (computer-aided design) and CIM (computer integrated manufacturing) application projects are two major projects approved by the State Council and promoted and implemented by the Ministry of Science and Technology of the PRC during the Eighth Five-Year Plan and Ninth Five-Year Plan periods. The application of CAD/CIM is oriented at the main battlefield of the Chinese economy. The manufacturing industry is transformed with information technology and modern management technology, and the competitiveness of enterprises is improved. During the Ninth Five-Year Plan period, the application and popularization of CAD/CIM made great progress. Its R&D has overcome a series of obstacles in major key technologies, established a group of high-level R&D and training bases, and reached the advanced international level in major key technologies such as overall technology, concurrent engineering, and integrated framework platform systems.

BGP Software R&D Center.

The network analyzer-computer-laser automatic repair system of the micro-assembly pilot lines of the 14th Research Institute of China Electronic Technology Group Corporation can be used to automatically test and repair hybrid integrated circuits such as receiving/sending modules.

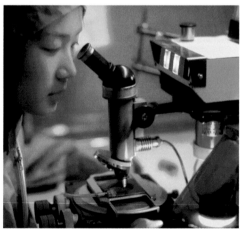

In the large-scale production test of integrated circuits, China has adopted the new self-invented simplified photo etching and polishing process. The illustration shows the photo etching exposure system of an integrated circuit.

Flexible manufacturing system workshop of Beijing First Machine Tool Works.

National CIMS Laboratory.

## The Sunway TaihuLight Supercomputer

Jin Yilian has led the development of a series of supercomputers, making significant contributions to China's seat in the world's high-performance computer field. In July 2000, China made a major breakthrough in developing and applying high-performance computers. Jiang Zemin named the high-performance computer system Sunway. Consequently, China became the third country in the world after the United States and Japan to be able to develop high-performance computers.

The National Meteorological Center used the Sunway Supercomputer to accurately run extremely complex mesoscale numerical weather forecasts, which played a key role in meteorological support for major events such as the 50th anniversary of the founding of the PRC and the return of Macao; the Shanghai Institute of Materia Medica of the CAS used it as a general drug research platform, which greatly shortened its development cycle of new drugs; The Institute of Atmospheric Physics of the CAS used it to perform parallel computing of the dynamic framework of the new generation of a high-resolution global atmospheric model, and has achieved encouraging results. The Sunway TaihuLight Supercomputer has been an indispensable high-end computing tool for meteorology, petroleum geophysical exploration, life science, aerospace, materials engineering, environmental science, basic science, and other fields. It has achieved significant benefits, playing an important role in China's economic construction and scientific research.

The remote-control mobile robot developed by the Shenyang Institute of Automation of the CAS can be used in hazardous environments such as the nuclear industry to replace manual labor on inspection, handling, equipment maintenance, and disassembly.

## Develop Intelligent Robots

The R&D of intelligent robots can bring substantial economic benefits and create new industries and employment opportunities. China has built a complete intelligent robot research system and developed many industrial robots as needed for future development. On November 30, 2000, the first humanoid robot independently developed in China with human appearance characteristics and the ability to simulate human walking and essential operation functions made its debut at the National University of Defense Technology. It is 1.4 meters tall, weighs 20 kilograms, possesses certain language functions, and walks two steps per second. It can move freely and walk in an uncertain environment with minor deviations. Conclusively, breakthroughs have been made in key technologies such as mechanical structure, control system structure, coordinated movement planning, control methods, etc.

The somatic-cell cloned goat.

## The First Somatic-Cell Cloned Goat in the World

Goat cloning was a key project of the NSFC and the Ministry of Agriculture chaired by Professor Zhang Yong. At the end of 1999, he successfully extracted two somatic cells from behind the ear of a small Shandong mountain goat, cultured them in vitro for a few days, and took out the nucleus. He took the oocyte of a goat just slaughtered, immediately denucleated it, injected the somatic nucleus into this oocyte, cultivated it into cloned embryos, and implanted them into the uterus of two white female goats, respectively. At 12:59 noon on June 16, 2000, one white female goat gave birth to a cloned goat. As it was the first somatic-cell cloned goat in the world, it is named Yuanyuan, meaning the first in Chinese. Unexpectedly, 36 hours later, Yuanyuan died of lung hypoplasia and hot weather. A week later, at 20:00 on June 22, its sister, another somatic-cell-cloned goat with the same appearance, Yangyang, was born. This marks that China's animal somatic-cell cloning technology has stood among the advanced ranks in the world, which has significantly impacted the development and improvement of its somatic-cell cloning technology.

## (4)  The Solid Scientific and Technological Foundation for Promoting Sustainable Social Development

### Achievements of Xia-Shang-Zhou Chronology Project

The Xia-Shang-Zhou Chronology Project is an unprecedented cultural project. As a national key scientific and technological project during the Ninth Five-Year Plan period, it was officially initiated on May 16, 1996, and passed the national inspection and acceptance on September 15, 2000. The project combined natural sciences with humanities and social sciences. It was the largest multidisciplinary systematic project in China. During its implementation, more than 200 experts and scholars from various disciplines, such as history, archaeology, astronomy, scientific and technological chronology, under the leadership of four chief scientists, Li Xueqin, Li Boqian, Xi Zezong, and Qiu Shihua, worked together, overcoming difficulties, and obtained innovative research results, completing the studies on nine general topics and 44 unique topics. The Xia-Shang-Zhou Chronology Project has given a chronology scale to the important period (Xia, Shang, and Zhou dynasties) of the development of Chinese civilization, clarified the evolution and development of the pre-Qin history, filled a gap in China's ancient chronology, and drawn the most scientifically based chronological table of the Xia and Shang Dynasties, laying a foundation for continuing to explore the origin of the Chinese civilization. Meanwhile, the successful completion of this project has also set the paradigm of interdisciplinary joint research in the 21st century.

Shangcheng, Zhengzhou, is the site of the early capital of the Shang Dynasty, with an area of about 13 square kilometers. The illustration shows the excavation site of the palace site on North Street in Shangcheng.

## Comfortable Housing in the 21st Century

China has begun implementing residential quarters' demonstration projects since the 1980s. It emphasized the improvement of supporting functions of single housing in the 1980s, proposed the complete supporting construction of residential quarters in the early 1990s, and initiated the National Comfortable Housing Demonstration Project at the end of the 1990s. In 1999, the Ministry of Construction issued the *Outline for the Implementation of the National Comfortable Housing Demonstration Project* and officially implemented it. Aimed at promoting the modernization of the housing industry, the project intended to drive the application of new processes, materials, equipment, and technologies in housing construction, improve housing design and construction level, the quality of living, and achieve comfortable housing.

It took science and technology as the guide, and promoting the modernization of the housing industry as the overall objective, to accelerate the transformation and updating of the traditional housing industry, which fully reflects the level of housing design in the 21st century, the design principle of "people-oriented, returning to nature." It has reached the standard of "healthy housing" (defined by the WHO), reflecting the scientific and technological content of building materials by fully applying "green building materials," thus leading the development direction of the Chinese housing industry.

After the Third Plenary Session of the Eleventh Central Committee of the CPC, Huaxi Village in the east of Jiangyin City, Jiangsu Province, has undergone tremendous changes, becoming the No. 1 Chinese Village famous at home and abroad.

The Comfortable Housing Project is a people-oriented project launched by the Ministry of Construction to promote the modernization of the Chinese housing industry and improve the quality of housing products, following the affordable housing project, well-off housing, and urban housing demonstration community.

# Full Implementation of the "Innovation and Industrialization" Policy—Implementation of the *Scientific and Technological Development Plan of the Tenth Five-Year Plan*

On May 18, 2001, the State Planning Commission and the Ministry of Science and Technology jointly released the *Special Plan for the Development of Science, Technology, and Education of the Tenth Five-Year Plan for National Economic and Social Development* (abbreviated as the *Scientific and Technological Development Plan of the Tenth Five-Year Plan*). Based on "orientation, reliance, and peak climbing," it proposed the guidelines of "doing something, but not everything, achieving overall follow-up, key breakthroughs, high-tech development, industrialization, improving the sustainable innovation capability of science and technology, and realizing technological leapfrog development," which is shortened as "innovation and industrialization." According to this policy, the strategic deployment was made at the two levels of "promoting industrial technology upgrading" and "improving the capability of sustainable scientific and technological innovation" in response to the urgent needs of the national economic development at that time and the strategic needs of the country's medium and long-term development.

## (1) National Science Plan Issued and Related Work

### Key Projects Plan of International Science and Technology Cooperation

Since the founding of the PRC, China has been engaged in extensive scientific and technological cooperation and exchanges with most countries in the world and has achieved fruitful results. In particular, the scientific and technological cooperation with the Soviet Union in the 1950s and with Western countries after the reform and opening up has played a prominent role in the history of China's foreign scientific and technological cooperation. Since the 1990s, there have been some new features and trends in global scientific and technological cooperation. Given domestic and international situations, China has formulated corresponding international cooperation policies and short-term goals. Their implementation has enabled its international scientific and technological cooperation projects to be conducted at a high level and share the achievements of major international cooperation projects mutually beneficially.

## Funds for Transformation of Agricultural Scientific and Technological Achievements

In 2001, with the approval of the State Council, the funds for the transformation of agricultural scientific and technological achievements were established to support the agricultural scientific and technological achievements to enter the early development of production and gradually set up a new type of agricultural scientific and technological investment guarantee system that adapts to the socialist market economy, conforms to the laws of agricultural development, and effectively supports the transformation of agricultural scientific and technological achievements into productive forces by attracting funds from enterprises, scientific and technological development institutions, financial institutions, and other channels.

## National Agricultural Science and Technology Park

The National Agricultural Science and Technology Park is a new model of agricultural development guided by the market and supported by science and technology. It is an integrated carrier of agricultural technology assembly, a link between the market and agriculture, a radiation resource of modern agricultural science and technology, a base for talent cultivation and technical training, and a demonstration and drive for the surrounding agricultural industry upgrading and rural economic development. The Ministry of Science and Technology initiated the National Agricultural Science and Technology Park pilot in 2001.

## Olympic Science and Technology (2008) Action Plan

This plan aimed to make the Beijing 2008 Olympic Games a sports event with the highest scientific and technological content. Seizing the opportunity of applying science and technology to improve the Olympics, it has enhanced China's scientific and technological innovation capability and the level of science and technology serving economic and social development and promoted its high-tech leapfrog development. Meanwhile, the Beijing 2008 Olympic Games made a window and platform to show the world China's high-tech innovation.

The successful development of the Dawning 4000A high-performance computer marks that the critical technologies of large-scale parallel computers in China have entered a new stage.

The second on the left is Li Guojie, chief designer of Dawning 4000A and academician of the CAE.

## (2) Concentrate Resources, Seize the Commanding Heights, and Achieve Leapfrog Development

To implement the strategy of invigorating China through science and education and the strategy of making China stronger through talents, and actively respond to the opportunities and challenges of the competition for skills, patents, and technical standards after China's accession to the WTO, the Ministry of Science and Technology, approved by the tenth meeting of the National Science and Education Steering Group in December 2001, organized the implementation of the "talent, patents, and technology" strategies of the Tenth Five-Year Plan, and launched 12 major science and technology projects. Through the organization and implementation of these major science and technology projects, we have tackled some long-term, fundamental, and overall strategic science and technology problems and promoted the development of high-tech enterprises, which was the top priority in the scientific and technological undertaking of the Tenth Five-Year Plan, a specific action to execute the important thought of Three Represents, and an objective need for scientific and technological work to keep pace with the times in the new age.

### Stronger Capability of Scientific and Technological Innovation and Achieve Leapfrog Development

In the *Scientific and Technological Development Plan of the Tenth Five-Year Plan*, China, from a strategic perspective, has offered stronger support for life science, nanoscience, information science, earth science, and other frontier fields and achieved many significant results. Its original innovation capability has been significantly enhanced, such as the breakthrough in plant gene sequencing, the successful completion of the pilot drilling task of CCSD-1, the Dawning 4000A high-performance computer ranking among the world's best, and its world-leading cloning and identification of 1,500 new genes related to human biological functions and diseases. In addition, the National High-tech Development Zone has realized extraordinary development, becoming an important technological innovation base in China; the national key laboratory in operation covers most disciplines of basic research and applied basic research in China.

*1) Dawning 4000A high-performance computer*

The computing capacity of Dawning 4000A high-performance computer developed by the Institute of Computing of the CAS has exceeded 10 trillion times per second, ranking 10th among the world's top 500 supercomputers selected by Lawrence Berkeley National Laboratory of the U.S. Department of Energy in June 2004. China has become the third country in the world after the United States and Japan that can manufacture and apply ten trillion-calculation-per-second commercial high-performance computers. Still, the CPU used by Dawning 4000A was produced by AMD in the United States.

*2) Completed 10% of the tasks of the international human genome HapMap (haplotype map) project*

Following the International Human Genome Project, the International Human Genome HapMap Project is another major research project in the field of genome research. It will provide complete human genome information and effective research instruments for the study of genetic polymorphism of different populations, analysis of disease and genetic association, determination of treatment genes and factors, analysis of efficacy, side effects, and disease risks, and research on the history of human origin, evolution, and migration. It will be the most powerful and economical tool for studying common human diseases. The program was started in October 2002, and China completed 10% (the HapMap of chromosomes 3, 21, and 8) of the whole international human HapMap with the best data quality in the world.

## Agricultural Science and Technology Have Promoted the Transformation of Agricultural Growth Mode and the Improvement of Comprehensive Productive Capacity

Because of the new trends and characteristics of the issues relating to agriculture, rural areas, and rural residents, according to the *Scientific and Technological Development Plan of the Tenth Five-Year Plan*, which is focused on the two prominent problems of increasing grain production and improving farmers' income, we have conducted the basic research on agricultural bio genomics and agricultural product quality improvement, sustainable control of agricultural pests, improvement of the ecological environment and efficient utilization of agricultural biological resources. The transformation, demonstration, and application of effective agricultural achievements have driven the adjustment and upgrading of agricultural structure, introduced and incubated clusters of scientific and technological leading enterprises, and promoted the transfer of the farm labor force and the increase of farmers' income. The Agricultural Science and Technology Popularization Center in

Shandong Zhucheng City Agricultural Science and Technology Promotion Center, with many precision instruments and equipment, accumulated many crop varieties and pest data files over the years.

many cities of Shangdong Province possesses a significant number of precision instruments and equipment. It has accumulated many crop varieties and pest data files over the years.

### 1) Quality improvement of agricultural products

The basic principles of agricultural science and technology work proposed in the Outline were implemented. The agricultural science and technology and education development plan of the Tenth Five-Year Plan was initiated to speed up the combination of traditional technology and high technology, the improvement of crop varieties and quality, and the improvement of breed and large-scale breeding in the breeding industry; to vigorously strengthen technological innovation in the field of agricultural product processing, the basic research and high-tech research of farming applications, and international cooperation and exchange in agricultural science and technology.

The germplasm resource bank of the Institute of Variety Resources is well-equipped and technologically advanced, standing among the world's advanced ranks.

*2) Sustainable control of agricultural pests*

Some progress has been made in the field trials and demonstrations of the Research on Sustainable Control Technology of Major Rice Pests in the South China Rice Region, a special topic of the Tenth Five-Year Plan Scientific and Technological Breakthrough Project. Through the field resistance variety evaluation test, new varieties that can be used for the assembly of sustainable control technologies were selected, such as Qilisimiao, Fengsizhan, Yuexiuzhan, etc.; we have carried out screening of pesticides with high efficiency and low toxicity and research on control technology, and developed the 58% Rice Pest Be Gone, which was listed as a designated insecticide for Taishan Zhenxiang organic rice (certified by the Ministry of Agriculture), and can replace the highly toxic pesticide methamidophos; we have performed the preliminary tests on the resistance of brown back rice plant hopper to imidacloprid and Buprofezin, studied the relationship between different planting densities of throwing seedlings and different fertilization modes and the occurrence of pests, so as to develop the relevant technology that adapts to high-quality, low nitrogen fertilizer, and wet-irrigation and that loosens the control indicator productive supporting technology measures for pest control, and to provide a basis for the research on sustainable control technology of significant rice pests in South China; based on the research on the critical technologies of sustainable control of major rice pests, we have proposed the sustainable control technologies with the key measures of disease resistant, high-quality and high-yield varieties, pest control and high-yield cultivation, and rational pesticide use.

*3) Protection and utilization of agricultural biological resources*

In accordance with the development trend of agricultural biological resources and environmental control during the Tenth Five-Year Plan period, we have promoted cooperation and collaboration among enterprises, universities, and research institutes under the guidance of the Tenth Five-Year Plan, accelerated the development of the bioenvironmental protection industry, improved its overall technical level and made contributions to sustainable agricultural development and pollution-free and green food production.

## Breakthroughs in Research on Key Common Industrial Technologies Have Promoted the Adjustment and Upgrading of Industrial Structure

The *Scientific and Technological Development Plan of the Tenth Five-Year Plan* has enabled China to have a good foundation for developing common and critical technologies in the industry and has conquered a number of key common industrial technologies, providing effective support for promoting industrial restructuring and technology upgrading. The launch of major projects has driven a group of large enterprise groups to significantly improve their independent innovation capabilities, such as the construction of the Three Gorges Project, the construction of the Qinghai–Xizang Railway, the transmission of gas from the west to the east, and the transmission of electricity from the west to the east. Through the special project of manufacturing informatization, we have mastered a series of key technologies for manufacturing informatization.

### *1) The Three Gorges Project*

On April 3, 1992, the Fifth Session of the Seventh National People's Congress reviewed and approved the *Resolution to Build the Three Gorges Project*. The Three Gorges Project is located in Sandouping, Yichang, Hubei Province, in the middle of Xiling Gorge, and its downstream is 40 kilometers away from Gezhouba Water Control Project. It controls a drainage area of 1 million square kilometers and has an average annual runoff of 451 billion cubic meters, thus creating comprehensive benefits such as flood control, power generation, and shipping. The Three Gorges Dam adopts the stage diversion mode.

The development of the Three Gorges Project will provide reliable and cheap power for the economically developed but energy-deficient central and eastern China and greatly improve the shipping conditions in the middle and lower reaches of the Yangtze River. In addition, it is conducive to promoting the development of fishery and tourism in the reservoir and implementing the South-to-North Water Diversion Project.

After more than ten years of hard work, the Three Gorges Dam was completed on May 20, 2006. The illustration shows its double-track five-level continuous ship lock.

Phase I of the project was completed on the mark that the Yangtze River was dammed on November 8, 1997; the second phase of the project realized the initial impoundment of the reservoir, the power generation of the first batch of generators, and the opening to navigation of its permanent ship locks. Its third phase achieves power generation of all generators and the improvement of the hub project. The total investment of the Three Gorges Project is more than 200 billion yuan, submerges 431,300 mu of cultivated land, and resettles 1,310,300 people in total. After completion, it has an average impound level of 175 meters, a flood control capacity of 22.15 billion cubic meters and a total water storage capacity of 39.3 billion cubic meters, making it the largest water control project in the world today.

### 2) West-East Natural Gas Transmission Project

Tarim, Qaidam, Shaanxi, Gansu, Ningxia, and Sichuan basins in western China contain 26 trillion cubic meters of natural gas resources, accounting for 87% of the onshore natural gas resources in China. In particular, the Tarim Basin in Xinjiang stores more than 8 trillion cubic meters of natural gas resources, accounting for 22% of the total Chinese natural gas resources. The discovery of natural gas in the Tarim Basin has made China a major natural gas owner after Russia, Qatar, and Saudi Arabia. In February 2000, the first meeting of the State Council approved the launch of the Project of Natural Gas Transmission from West to East China, another major investment project second only to the Three Gorges Project and a landmark construction project to kick off China's Western Development. The planned Project of Natural Gas Transmission from West to East China covers the Central Plains, East China, and the Yangtze River Delta. It starts from Lunnan Oil and Gas

The main body of Asia's largest natural gas processing plant has been completed.

As the backbone project of China Western Development, the West-to-East Power Transmission Project develops the power resources of Guizhou, Sichuan, Inner Mongolia, and other western provinces and regions and delivers electricity to Guangdong, Shanghai, and Beijing-Tianjin-Tangshan regions where there is a shortage of electricity.

Field in Tarim, Xinjiang in the west, passes through Korla, Turpan, Hami, Jiuquan, Zhangye, Wuwei, Lanzhou, Dingxi, Xi'an, Luoyang, Xinyang, Hefei, Nanjing, Changzhou and other cities in the east, and ends in Shanghai. It runs through 9 provinces (cities and districts) in Xinjiang, Gansu, Ningxia, Shaanxi, Shanxi, Henan, Anhui, Jiangsu, and Shanghai from east to west. The implementation of this project is conducive to promoting the adjustment of the Chinese energy structure and industrial structure, driving the economic development of the East and West, improving the quality of life of the people in the Yangtze River Delta and the areas along the pipelines, and effectively controlling air pollution. It has created conditions for China's Western Development and the transformation of the resource advantages of the western region into economic benefits. It is of great strategic significance to promote and accelerate Xinjiang's economic development and the west's territory.

### 3) West–East Power Transmission Project

The uneven distribution of energy resources in China is not only that natural gas is primarily distributed in the central and western regions, but also oil, coal, and hydropower. After years of exploitation, most of the coal mines and oil fields in the eastern region are depleting, and energy shortage has become a major problem. To alleviate the problem of an energy shortage, in addition to the Project of Natural Gas Transmission from West to East China, the West-to-East Power Transmission Project is also an important measure, safe, reliable, clean, and cheap. Therefore, the Tenth Five-Year Plan regards it as one of the key construction projects and a landmark project of China's Western Development.

This project comprises three channels: the South, the Middle, and the North. The South channel refers to hydropower development in the southwest, thermal power in Yunnan and Guizhou, and power transmission to Guangdong. The middle channel refers to extending the power transmission network westward to the upper reaches of the Yangtze River, with the power transmission from the Three Gorges as the leader, to realize the joint power transmission from Sichuan, Chongqing, and Central China to East China and Guangdong. The North channel refers to the gradual realization of power transmission from hydropower in the upper reaches of the Yellow River and thermal power in the Three Xi (Hexi, Dingxi, Xihaigu) region to North China and Shandong based on power transmission from northern Shanxi and western Inner Mongolia to the Beijing-Tianjin-Tangshan region.

In 2000, the first batch of construction in Guizhou and Yunnan of the West-to-East Power Transmission Project started, marking the full launch of the West-to-East Power Transmission Project in China. During the Tenth Five-Year Plan period, the second batch of construction was started. As the generators of the Three Gorges Dam were put into operation in succession, additional 4.2 million kilowatts of power were transmitted to East China and extra 3 million kilowatts of power to Guangdong in the middle channel. The transmission projects from the Three Gorges to East China and from Sichuan to East China have been completed, realizing the AC networking between the Sichuan-Chongqing power grid and the Central China power grid. The North Channel added 5 million kilowatts of power to Beijing, Tianjin, Tangshan, and Hebei South Power Grid. The transmission of power from west China to east China is of far-reaching significance to improving the Chinese power structure. It has directly boosted the development of electric power in the western region, which is of great significance to the adjustment of the national electric power structure and the optimization of its distribution.

## Deployment in Advance for Major Scientific and Technological Issues

The *Scientific and Technological Development Plan of the Tenth Five-Year Plan* has advanced energy and resource R&D deployment at different levels in China and has achieved specific results. For example, the 10 MW high-temperature gas-cooled experimental reactor has completed 72 hours of full power generation operation. With the breakthrough and industrialization of clean coal technology, coal water slurry gasification, and other technologies, the energy conservation level has been dramatically improved; major scientific research, such as the formation mechanism of major disasters and major disaster reduction projects, provides strong technical support for ecological research and construction planning and disaster prevention and reduction; we have conquered a number of key technologies and standards to ensure people's health and social stability and development. For example, the completion of clinical research on avian influenza vaccine for human use marks a major breakthrough in this field; focused on the four steps of "check, trace, limit, and control," the food safety project researched food safety standards, control technologies, and key detection, establishing the first food pollution monitoring network covering 13 provinces and cities in China.

### 1) Avian influenza vaccine for human use

As a key scientific and technological project of the Tenth Five-Year Plan in China, a Phase II clinical trial of the R&D of avian influenza vaccine for human use was conducted from September to November 2007. A total of 402 subjects aged between 18 and 60 participated in the trial. The results showed that the vaccines of three antigen doses used in clinical trials could induce the human body to produce antibodies to a certain extent. Among the vaccines, all three indicators, the protective antibody positive rate, antibody positive conversion rate, and geometric mean titer (GMT) multiple of 10 micrograms and 15 micrograms, reached the internationally recognized vaccine evaluation standard, indicating that the vaccine has good immunogenicity to the human body. No serious adverse reactions occurred from the observation results of the subjects' local and systemic adverse reactions, indicating that the vaccine is excellently safe. In 2009, the clinical trial was completed, and its mass production was approved.

The illustration shows the control room of the HTR 10 with full digital control and protection system.

On June 4, 2009, the LAMOST at the Xinglong Observatory of the NAOC of the CAS passed the national inspection and acceptance.

## 2) HTR 10

The 10 MW high-temperature reactor (HTR 10) built by Tsinghua University is the world's first modular pebble-bed high-temperature gas-cooled reactor and one of the significant projects in the energy field of Project 863 in 2000. A high-temperature gas-cooled reactor (HTGR) is an advanced nuclear technology with the main technical characteristics of fourth-generation nuclear power because of its inherent safety features, high temperature, and wide applications. It can be used as a supplement to large, pressurized water reactor (PWR) nuclear power plants. Together, they meet the country's strategic needs to develop nuclear power actively. This pebble-bed HTR 10, using helium as a coolant, graphite as a moderator, and structural material, can withstand a high-temperature range of 700–1,000°C .It provides high-temperature process heat by taking advantage of its high outlet temperature, thus meeting the demand for high-temperature process heat in petroleum thermal recovery, steelmaking, chemical industry, coal gasification, liquefaction, etc. Large-scale hydrogen production can replace imported liquid fuel, a new field of nuclear energy utilization, and one of the main purposes of developing advanced nuclear technology, especially HTGR.

## 3) Large Sky Area Multi-Object Fiber Spectroscopic Telescope (LAMOST)

The National Astronomical Observatories of China (NAOC)'s LAMOST is a giant optical telescope with independent Chinese innovation and a Schmitt telescope with transit-instrument-style reflection, lying in the north-south direction. Its application of active optics to control the reflection correction plate makes it the world's largest optical telescope with a large aperture and a large field of view. The 235-million-yuan super telescope, with an aperture of 4 meters, can observe objects up to an opacity of 20.5 within 1.5 hours of exposure. Because of its 5-degree field of view, 4,000 optical fibers can be placed on the focal plane to transmit the light of distant celestial bodies to multiple spectrometers and obtain their spectra simultaneously, making it the telescope with the highest spectral acquisition rate in the world. This telescope is installed in the Xinglong Observatory of the NAOC and is an internationally leading large scientific device in large-scale optical spectrum observation and large field astronomy research in China. The LAMOST project is divided into seven subsystems: optical system, active optics and support system, rack and tracking device, telescope control system, focal plane instrument, dome, data processing, and computer integration. The telescope project was started in 2001 and completed in 2008.

The first maglev train pilot line in China, developed by Southwest Jiaotong University, has a total length of 43 meters. When running, the train can be suspended in the guideway about 8 mm at 30 kilometers per hour. Its completion has helped China further to strengthen the research of superconducting technology and maglev train,

*4) Research on high-speed maglev transportation technology*

As early as the 1970s, China began its research on the application of maglev transportation technology. On September 29, 2005, the CM1 Dolphin high-speed maglev vehicle component, a high-tech project that has attracted wide public attention, part of Project 863, was officially produced in Chengdu Aircraft Industrial (Group) Co., Ltd. High-speed maglev transportation technology is one of the major projects in the transportation field of the National Science and Technology Support Plan. It mainly includes R&D of a full set of technologies, equipment, and components, such as 500 km/h high-speed maglev vehicles, levitation guidance control technology, traction control technology, operation control technology, and system integration technology, the establishment of high-speed maglev transportation system planning, design technology, and standard system, construction of a 30 km high-speed maglev train pilot line, and the completion of the finalized industrial test with independent intellectual property rights.

*5) Successful launch of Shenzhou 5*

China's millennium dream has been to explore space and traverse the universe. On October 15, 2003, the Shenzhou 5 manned spacecraft was successfully launched. As China's first astronaut to go to space, Yang Liwei returned to Earth safely after 14 circles of flight around the Earth. The smooth return of the Shenzhou 5 spacecraft marks that China has become the third country to complete a manned spacecraft space flight after the United States and the Soviet Union.

The Shenzhou 5 manned spacecraft was successfully launched. The illustration shows Yang Liwei preparing to go on the space expedition.

## (3) Actively Conduct High-Level International Exchanges

### 2002 International Congress of Mathematicians (ICM)

On August 20, 2002, the 24th ICM held a grand opening ceremony in Beijing's Great Hall of the People. President Jiang Zemin attended the event. With a history of more than 100 years, the ICM is the highest-level global academic conference on mathematical science. Its first meeting was held in Zurich, Switzerland, in 1897, and it has been held every four years since the Paris ICM in 1900. Except during the two world wars, it was never interrupted. The 24th ICM was the first time that it was held in a developing country in its history. It was the first ICM in the 21st century and the largest ICM in history. A total of 4,157 mathematicians from 104 countries and regions attended the conference, including 1,965 mathematicians from mainland China. A total of 20 mathematicians were invited to make a report, which fully reflects the mutual penetration and connection of different mathematical fields and the deeper intersection of mathematics and other sciences.

In addition to reports, the ICM organized a series of forums and activities for the public and teenagers, such as three public reports, 46 satellite conferences, and the Walk into the Wonderful Math Garden Youth Math Forum. A great number of top mathematicians from all over the world gathering in Beijing has created an excellent opportunity for Chinese mathematicians to learn from, exchange, and discuss with international masters. It also provided a good platform for Chinese mathematicians to present their work to international peers. The conference and related activities have great significance and far-reaching influence on the promotion and popularization of mathematics, the enhancement of social attention to mathematics, the promotion of the application of mathematics in all walks of life in China, the strengthening of academic exchanges between Chinese mathematicians and international peers, and the promotion of global mathematics into a new age.

From August 20 to 28, the 24th ICM was held in Beijing.

## The 28th General Assembly of the International Council of Science Unions (ICSU)

From October 18 to 22, 2005, the 28th General Assembly of the ICSU was held in Suzhou. ICSU is the most authoritative non-governmental international organization in the scientific community. Founded in 1931, it is known as the "United Nations of the scientific community." Gathering the representatives of various major fields of natural science, its academic activities can basically represent the level and trend of scientific development in the present world. The 28th General Assembly of the ICSU gathered more than 270 first-class scientists from 64 countries and regions (including several Nobel laureates), more than ten academicians, and more than 300 experts and scholars from China who participated in support. Around 600 Chinese and foreign scientists attended the event. The opening ceremony was hosted by Lubchenko, president of the ICSU. State Councilor Chen Zhili attended the opening ceremony and delivered a speech on behalf of the Chinese government. Lu Yongxiang, Vice Chairman of the Standing Committee of the National People's Congress and president of the CAS; Zhou Guangzhao, Chairman of the CAST; Deng Nan, Secretary of the Party Leadership Group of the CAST; Wu Zhongze, Member of the Party Leadership Group of the Ministry of Science and Technology; Chen Zhu, Vice President of the CAS; Sun Honglie, Chairman of the Chinese Committee of the ICSU; Hu Qiheng and Wei Yu, vice chairmen of the CAST; and Feng Changgen and Cheng Donghong, secretaries of the Secretariat of the CAST attended the conference. As the conference's host country, China sent a delegation of 20 famous domestic scientists, including Zhou Guangzhao, Lu Yongxiang, Sun Honglie, Hu Qiheng, and Wei Yu, to attend the conference. The 28th General Assembly of the ICSU was an unforgettable international scientific event.

In October 2005, the 28th ICSU Congress was held in Suzhou.

# Make China Innovative

V

On January 9, 2006, the National Science and Technology Conference was held in the Great Hall of the People in Beijing, which was the first national science and technology conference in the 21st century. The illustration shows the venue of the conference.

On January 9, 2006, the Central Committee of the CPC and the State Council convened the National Conference on Science and Technology, making a major strategic decision to make China innovative. To build an innovative country, the core is to take strengthening the capability of independent innovation as the strategic basis for the development of science and technology, so as to walk the path of independent innovation with Chinese characteristics, and promote the leapfrog development of science and technology; it is to take strengthening the capability of independent innovation as the central link in adjusting the industrial structure and transforming the growth mode, so as to build a resource conserving and environment-friendly society, and promote the rapid and sound development of the Chinese economy; it is to take strengthening the capability of independent innovation as a national strategy, running it through all aspects of modernization, so as to stimulate the innovation spirit of the whole nation, train high-level innovative talents, form systems and mechanisms conducive to independent innovation, vigorously promote theoretical innovation, institutional innovation, scientific and technological innovation, and constantly consolidate and develop the great cause of socialism with Chinese characteristics.

# Independent Innovation, Key Leaps, Supporting Development, Leading the Future— Implementation of the *National Medium and Long-Term Plan for Science and Technology Development (2006–2020)*

On February 9, 2006, the State Council issued the *Outline of the National Medium and Long-Term Plan for Science and Technology Development (2006–2020)* (hereinafter referred to as the *Plan Outline*). Its formulation was a major task proposed by the 16th National Congress of the CPC and an important measure to make China innovative. The *Plan Outline* points out that in the next 15 years, the guiding principle of a scientific and technological undertaking is: independent innovation, key leaps, supporting development, and leading the future. Independent innovation strengthens the original innovation, integrated innovation, introduction, understanding, absorption, and re-innovation from the perspective of enhancing the national innovation capacity. The key leap is to stick to doing something. Still, not everything; select critical areas with specific foundations and advantages related to the national economy and people's livelihood and national security, and focus on breakthroughs to achieve leapfrogging development. Supporting development starts with the urgent actual needs, makes efforts to break through significant key and standard technologies, and supports the sustainable and coordinated development of the economy and society. Leading the future is taking a long-term view, planning on cutting-edge technology and basic research in advance, creating new market demand, cultivating emerging industries, and leading future economic and social development.

Shenyang Blower Works has basically realized the automatic design process by adopting CIMS technology.

# (1) Establish Science and Technology Innovation System

## Accelerate the Advancement of Science and Technology

Since the 21st century, economic globalization has picked up the pace significantly, and the momentum of the world's new scientific and technological revolution has become stronger. A series of recent major scientific discoveries and technical inventions are being transformed into real productive forces at a faster rate, profoundly changing the face of the economy and society. The leading role of science and technology in promoting economic development, fueling social progress, and safeguarding national security has become more prominent, and international competition based on scientific and technological innovation has become more fierce. Major countries in the world regard scientific and technological innovation as an important national strategy, scientific and technological investment as strategic input, and the development of strategic technologies and industries as an important breakthrough to achieve leaps and bounds. In the long run, China must have a national science and technology innovation system based on the whole nation. This system should go beyond all departments and units and include defense science and technology innovation to a large extent. That is, if the defense system is not linked with the innovation system of the whole nation, it cannot organize and mobilize the wisdom of the entire country, which will hinder the growth and prosperity of the national defense capability.

## Improve the Independent Innovation Capability of Enterprises

The 17th National Congress of the CPC report proposed: "We should adhere to the path of independent innovation with Chinese characteristics and enhance the capability of independent innovation in all aspects of modernization." Therefore, it has become an important subject that we must seriously study and explore to focus on improving the independent innovation capability of enterprises and further transforming the mode of economic growth. The so-called independent innovation capability refers to the capability to continuously promote the innovation of economic structure and sustainable economic growth under the guidance of the scientific development concept. It starts with enhancing our innovation capability, taking our strength as the main body, applying innovative knowledge, new technologies, and processes, and adopting new production methods and management models.

## Establish and Improve the Intellectual Property Protection System

In the process of economic globalization, intellectual property, as an important guarantee for the development of the knowledge economy, has become the source of industrial core competitiveness. In the age of the knowledge economy, due to the need for a sound intellectual property management system, the improvement of the core competitiveness and strategic advantages of enterprises has been seriously challenged. China has built a comprehensive system of intellectual property laws and regulations and law enforcement and protection systems that adapt to Chinese conditions, conform to international rules, and cover a wide range of areas. Under this law enforcement protection system, China has adopted a distinctive intellectual property protection mode of "two approaches in parallel operation" of judicial protection and administrative protection.

Saddle-shaped superconducting magnet for coal-fired magnetofluid power generation developed by the Institute of Electrical Engineering, CAS.

## (2)  Create an Environment for Independent Innovation

### Promote Technological Innovation of Enterprises

On February 26, 2006, the State Council issued *Several Supporting Policies for the Plan Outline* (hereinafter referred to as Supporting Policies). The *Supporting Policies* revolve around three main links: increasing the input of innovation elements, improving the efficiency of innovation activities, promoting the realization of innovation value, creating an encouraging new environment for independent innovation, promoting enterprises to become the main body of technological innovation, strives to build an innovative China, and implements the policies in ten aspects, which are respectively science and technology investment, tax incentives, financial support, government procurement, introduction, understanding, absorption and re-innovation, creation and protection of intellectual property rights, talent team, education and science popularization, scientific and technological innovation base and platform, and stronger overall coordination.

### Implement Innovative Talent Training Projects

The *Supporting Policies* pointed out that China would implement the national high-level innovation talent training project and cultivate a group of high-level discipline leaders with strong innovation capability in a number of strategic science and technology fields related to national competitiveness and security, such as basic research, high-tech research, social welfare research, and form excellent innovation talent clusters and innovation teams with Chinese characteristics. They aim to improve and perfect the academic exchange system, enhance the peer recognition mechanism, and enable young and middle-aged outstanding scientific and technological talents to stand out; to establish a talent evaluation and reward system that is conducive to encouraging independent innovation; to reform and improve the distribution and incentive mechanism for enterprises, support them in attracting scientific and technological talents, and allow state-owned high-tech enterprises to implement incentive policies such as offering options for technical and managerial backbones.

### University Science Parks

In November 2006, the Ministry of Science and Technology and the Ministry of Education issued the *Measures for the Identification and Management of National University Science Parks*. It mainly provides macro management and guidance for university science parks and points out that universities are the main supporting units for developing national university parks. The National University Science Park is an important part of the national innovation system and an important base for independent innovation. It is one of the important platforms for universities to realize the combination of production, education, research, and social service functions. It is also one of the primary sources of innovation for high-tech industrialization, for the "starting a new undertaking" of a national high-tech industry development zone, and for promoting regional economic development and supporting industrial technological progress. It is also a part of the higher education system with Chinese characteristics. The University Science Park is one of the important symbols of a first-class university. It should establish a management system and operating mechanism that adapt to the socialist market economy, improve the park infrastructure construction, service support system construction, industrialization technical support platform construction, college student internship and practice base construction through various ways, and provide all-round and high-quality services for entrepreneurs in the park.

On November 6, 2007, people visited the new generation electric vehicle model platform exhibited at Tongji University. At the 2007 China International Industry Expo, which opened in Shanghai on the same day, the university exhibition zone was particularly eye-catching, with more than 50 universities displaying hundreds of scientific and technological innovations and achievements of industry-university-research cooperation.

Tianjin University Science Park is one of the first 15 university science parks explicitly approved by the Ministry of Science and Technology and the Ministry of Education. It covers an area of 131,600 square meters, with a building area of 100,982 square meters. The project cost 312 million yuan to build.

The agricultural science and technology park attaches importance to economic benefits and uses modern supporting cultivation technology to realize the modern agricultural model of high input and high output. Nowadays, vegetable greenhouses are free from the seasons. They are always full of fresh flowers and green trees, fragrant all year round. The illustration shows a partition fence composed of flowers, which adds color to the vegetable greenhouses.

## Agricultural Science Park

The agricultural science park is a new model of agricultural development oriented by the market and supported by science and technology. Being the carrier of agricultural technology assembly and integration, the link between the market and farmers, the radiation source of modern agricultural science and technology, and the base for talent training and technical training, it plays a demonstration and promotion role in the agricultural industry upgrading and rural economic development in the neighboring areas. Of a certain scale, it has feasible overall, clear leading industry, reasonable functional zoning, and significant comprehensive benefits. It has strong scientific and technological development ability, a relatively complete talent training, technical training, technical service and promotion system, and strong scientific and technological investment. It generates significant economic, ecological, and social benefits and plays a strong guiding and demonstration role for the neighboring areas. It has standardized land, capital, talent, and other rules and management systems and an operating mechanism that conforms to the laws of the market economy, is conducive to introducing technology and talents, and constantly expands investment and financing channels.

## Industrial Science Park

The Industrial Park, with multiple functions, can attract domestic and foreign investment and technologies, facilitate international cooperation, and become a part of the unified global market. It mainly serves start-ups, mature enterprises, investors, and small and medium-sized business operators, whose needs differ from residential customers. They mainly demand a business environment, such as the atmosphere and maturity of the park, whether it is convenient for enterprises to operate, and whether they can survive, develop, and establish a good corporate image. The vast majority of science parks house a number of start-ups that require venture capital support in "incubators." One of the characteristics of start-ups is that there are research projects and professionals engaged in development, but there is a lack of business management personnel. Nobody is handling the company's registration or building an effective operating mechanism. This has created a unique customer demand market for science parks.

Shekou Industrial Zone was independently developed by China Merchants Group Hong Kong in Shenzhen in 1979. After 20 years of development and construction, Shekou Industrial Zone has become a coastal city with full-equipped investment environment, complete service functions, and a beautiful living environment.

# Full Implementation of the Scientific Outlook on Development to Ensure the Fundamental Interests of the Chinese People—Implementation of the *Development Plan for Science and Technology during the Eleventh Five-Year Plan Period*

In the 21st century, scientific and technological innovation has become the main driving force for economic and social development. The Eleventh Five-Year Plan period is a critical period for China to implement the scientific outlook on development comprehensively, take strengthening its capacity for independent innovation as a national strategy, accelerate the transformation of economic mode, promote the optimization and upgrading of industrial structure, and lay the foundation for building a well-off society in an all-round way. The *Development Plan for Science and Technology during the Eleventh Five-Year Plan Period* defines the development ideas, goals, and priorities for the next five years according to the tasks and requirements defined in the *Plan Outline*, vigorously promotes scientific and technological progress and innovation, and lays a solid foundation for building a new China. Comrade Hu Jintao put forward in the report of the 17th National Congress of the CPC that the Scientific Outlook on Development is a major strategic outlook that, based on the basic national conditions of the primary stage of socialism, summarizes China's development practice, draws on foreign development experience and adapts to new development requirements. He emphasized that recognizing the basic national conditions in the primary stage of socialism does not mean belittling oneself and surrendering to backwardness, nor being divorced from reality and eager for success, but taking it as the fundamental basis for promoting reform and planning development. We must always stay sober, base ourselves on the greatest reality of the primary stage of socialism, make scientific analysis, deeply grasp the new topics and contradictions facing China's development, more consciously follow the path of scientific development, and strive to open up broader prospects for the development of socialism with Chinese characteristics.

# (1) Aim at Strategic Objectives and Implement Major Special Projects

China has executed the overall arrangement of the *Plan Outline*, focused on the combination with significant national projects, and coordinated and interacted with the national science and technology plan arrangements. It has given full play to the primary role of the market in allocating resources, formed a diversified input mechanism while ensuring the central financial input, and highlighted the main function of enterprises in technological innovation. During the Eleventh Five-Year Plan period, China has achieved full implementation and phased results of 16 major special science and technology projects, including core electronic devices, high-end general-purpose chips, basic software products, large-scale manufacturing equipment of integrated circuit and complete processes, a new generation of broadband wireless mobile communication network, high-end CNC machine tools, and basic manufacturing equipment, large-scale oil and gas field and coalbed methane development, large-scale advanced pressurized water reactor and high-temperature gas-cooled reactor nuclear power plants, water pollution control and treatment, new varieties of genetically modified organisms cultivation, major new drug development, prevention and treatment of major infectious diseases such as AIDS and viral hepatitis, large-scale aircraft, high-resolution earth observation systems, manned space flight, and lunar exploration projects, etc.

## High-End Manufacturing Equipment of Integrated Circuit

The 65-nanometer dielectric etcher specially designed for the integrated circuit equipment special project has been tested nearly 100 times by multinational customers. Compared with the chip processing results of the most advanced equipment in the world, it has excellent processing quality, 35%–50% higher unit investment output, and 30%–35% lower cost. In 2010, 12 sets of etchers were sold, and foreign batch orders were obtained, which significantly improved the international competitiveness of the Chinese high-end manufacturing equipment of the integrated circuit industry and drove the development of a series of emerging industries such as solar energy and flat panel display.

## The Chinese Fourth Generation Mobile Communication as the International Standard

At the beginning of 2008, the Executive Meeting of the State Council reviewed and approved the implementation plans for major national special projects for the first time, including the new generation broadband wireless mobile communication network. This time, the new generation broadband wireless mobile communication has once again become one of the 11 major science and technology projects whose implementation was to be sped up. It represents the main development direction of information technology. The implementation of this special project has dramatically enhanced the comprehensive competitive strength and innovation capability of China's wireless mobile communication and promoted China's mobile communication technology and industry to leapfrog to the world's advanced level.

Project 863 focused on upgrading domestic computers with technologies such as intelligent networks and high-performance computers.

The network control center, which resides on the bottom floor of the Cyberport office building, is the heart of Cyberport. The staff must pass both the fingerprint and pupil monitoring systems to enter the building.

On January 18, 2012, TD-LTE, created in China, was determined by the International Telecommunication Union as the international standard of fourth-generation mobile communication (4G). TD-LTE is an upgrade and evolution technology of TD-SCDMA (China Mobile 3G system) with independent intellectual property rights. Under the guidance of the Chinese government and the strong promotion of China Mobile, TD-LTE has received extensive attention and support from international operators.

## Preliminary Establishment of Transgenic Technology System

During the Eleventh Five-Year Plan period, 36 new insect-resistant cotton varieties were cultivated by China's special project for breeding new transgenic varieties. They were promoted to 167 million mu of land in three years, with a net benefit of 16 billion yuan. The insect-resistant transgenic rice and phytase transgenic maize have obtained safety production certificates. In front of the public attention, China has initially built its transgenic technology system and safety evaluation technology system.

## Progress in New Drug Development

With the support of the special project of new drug development, a batch of new drugs with independent intellectual property rights have been developed one after another. Among them, 16 products have obtained new drug certificates, 24 varieties have submitted new drug registration applications, 17 varieties, including the anti-lung cancer drug Icotinib, have completed clinical trials, and 36 large varieties of drugs have successfully implemented technical transformation.

## Prevention and Treatment of Major Infectious Diseases Such as AIDS and Viral Hepatitis

Focused on making breakthroughs in key technologies such as the development of new vaccines and therapeutic drugs, China has independently developed 40 kinds of highly effective and specific diagnostic reagents, 15 kinds of vaccines and drugs, studied and formulated scientific and standardized prevention and treatment plans of Chinese medicine, western medicine, and their combinations, established ten prevention and treatment technology platforms as good as those in developed countries, and initially built a technical system for effective prevention and control of AIDS and viral hepatitis.

AIDS is one of the serious problems threatening human survival and social development in the world today. Declaring war on AIDS is the most arduous chapter in the history of the human struggle against disease, and AIDS prevention and control brook no delay. The illustration shows researchers separating AIDS detection reagents.

The business of PetroChina Company Limited is divided into exploration and production, refining and sales, chemical industry and sales, and natural gas and pipeline. The illustration shows the rapidly developing oil and gas pipeline network. A pipeline can transport oil and gas directly to the wharf from the place of origin or processing sites.

## Oil and Gas Exploration and Development and Enhanced Oil Recovery Efficiency

In the field of energy, oil and gas development in China has conquered a number of core key technologies, such as oil and gas exploration and development and enhanced oil recovery efficiency. The first seismic data acquisition and recording system with a magnitude of more than 10,000 channels in China has successfully passed the actual production assessment of 2,000 engineering prototypes. Its high-speed data transmission capacity is 2 to 5 times higher than the current international mainstream products. The successful development of a 3,000 m deep-water semi-submersible drilling platform has enabled China's oil and gas industry to achieve a leapfrog development from a depth of 500 m to 3,000 m.

China has conducted focused research on high-precision seismic exploration and production technology for oil and gas, coalbed methane, and deep-sea oil and gas resources under complex geological conditions in western China, improved the independent design and manufacturing capacity of complete sets of technologies and equipment, and increased the exploration rate of oil and gas resources by 10% and 20% respectively, and the oil recovery efficiency by 40%–50%.

## Successful Development of Various CNC Machine Tools

The heavy five-axis-combined turning and milling machine tool and super heavy horizontal boring lathe successfully developed in the numerical control machine tool special project have played an important role in the construction of the third-generation nuclear power autonomy. The world's largest 36,000-ton ferrous metal vertical extruder has been put into production, and it can save more than 10 billion yuan of costs for relevant enterprises every year.

## Water Pollution Control and Treatment

Regarding water pollution control and treatment, China has selected different typical watersheds, carried out their water ecological function zoning, studied key technologies for their water pollution control, lake eutrophication prevention and control, and water environment ecological restoration, broken through drinking water source protection and drinking water advanced treatment and transmission technologies, developed safe drinking water assurance integration technology and water quality and quantity optimization and allocation technology, and established a technical system for water pollution monitoring, control, and water environment quality improvement that is suitable for the national conditions of China. Around "three rivers, three lakes, one river, and one reservoir," the major special project for water pollution control and treatment focused on tackling a number of critical technologies for water pollution prevention that need to be urgently addressed for energy conservation and emission reduction, to provide strong support for achieving the goal of energy conservation and emission reduction and improving the quality of water environment in key watersheds.

In the pollution control of Dianchi Lake, the bed mud dredging project (Phase I) has been completed ahead of schedule, which is the largest lake improvement project in China so far.

## (2) Face Urgent Needs and Conquer Key Technologies

During the Eleventh Five-Year Plan period, China, based on the urgent needs of national economic and social development, highlighted the development of major technologies in key areas such as energy, resources, environment, agriculture, information, and health, strengthened the R&D of public welfare technologies and generic industrial technologies, focused on integrated innovation and application centered on primary products and pillar industries, combined with significant project construction and major equipment development, intensified integrated innovation, introduction, understanding, absorption, and re-innovation, and improved the independent R&D capability of major industries in China.

### Major Breakthrough in Wide Area Real-Time Precision Positioning Technology

The wide area real-time precision positioning technology and demonstration system is a key project in the field of earth observation and navigation technology in Project 863 during the Eleventh Five-Year Plan period. It aimed to build a high-precision satellite navigation augmentation demonstration system in China based on wide area differential and precise single point positioning technology, by entirely using China's existing satellite navigation ground reference station resources and integrating advanced real-time data processing, Internet, and satellite communication technologies. During the Eleventh Five-Year Plan period, a satellite navigation augmentation demonstration system that entirely covers China was built, with real-time positioning accuracy of better than 1 meter, navigation satellite real-time orbit determination accuracy of 0.1 meters, and clock correction of 0.2 nanoseconds. Its main technical indicators have reached the advanced international level, marking that China has made a major breakthrough in the field of wide-area real-time precision positioning technology. The results of this project can be directly applied to the construction of the Beidou satellite navigation augmentation system in China and help significantly improve the positioning accuracy of the Beidou system.

### Continuous Stable Production above 40 Million Tons in the Late Stage of High Water Cut in Daqing Oilfield

Relying on the third-generation independent innovation technology, Daqing Oilfield had 56 million tons in 1995, achieving an annual output exceeding 50 million tons for 20 consecutive years. Next, the oilfield has entered a production decline stage. With the resources and technologies at that time, by 2009, the annual output had dropped below 30 million tons. To meet the country's urgent demand for oil, Daqing Oilfield has independently developed the matching technology of quantitative description of highly dispersed remaining oil and fine oil recovery, pioneered the polymer viscoelastic oil displacement theory and the efficient development technology of polymer flooding, first revealed the oil and gas distribution law of the negative structural belt in the large continental depression basin, innovated the fine oil exploration technology for thin sand bodies, and innovated and implemented the matching technology for the treatment and utilization of super large capacity diversified injection and production fluids, thus achieving a high and stable annual yield of over 40 million tons in 14 years.

## (3) Grasp Future Development and Make Plans about Cutting-Edge Technology and Basic Research in Advance

During the Eleventh Five-Year Plan period, investment in basic research has increased significantly, thus developing a number of emerging inter-disciplines and solving plenty of critical scientific problems in national economic and social development. China has maintained its comparative advantage in the field of aerospace technology and won the initiative in strategic areas such as information, biology, new materials, and oceans, striving to achieve a batch of original innovation achievements that reach the world's advanced level at the junction of the country's significant needs for future development and cutting-edge technologies and to form a group of technology systems and products that represent the world's advanced level.

### Development of Insight

The first special astronomical satellite independently developed by China, Insight, a hard X-ray modulation telescope (HXMT) included in the *Space Science Development Program during the Eleventh Five-Year Plan Period*, is a cube-shaped satellite with a total mass of about 2.5 tons. It is loaded with four detection payloads: high-energy, medium-energy, and low-energy X-ray telescopes and a space environment monitor. Through three working modes of sky observation, fixed-point observation, and small sky scanning, X-ray space observation with wide band, high sensitivity, and high resolution can be accomplished. Insight was launched on June 15, 2017, and has participated in international space and ground joint surveys many times. Through Insight, human beings have "heard" and "seen" the violent explosion in the deep universe for the first time and directly detected the "roar" by the merger of two neutron stars; the hard X-ray cyclotron absorption line observation of the strongest magnetic field in the universe found the "missing link" of the internal structure of the neutron star, providing new clues for studying the ultrahigh density, superfluid, and superconductor structure of the neutron star. In the future, Insight will provide Chinese scientists with highly sensitive images of celestial bodies (black holes), neutron stars, and neutron star binaries in the distant universe.

### Microgravity Science

Microgravity science studies the laws of material movement in a microgravity environment. Many macro motion processes in nature inevitably succumb to gravity in the terrestrial environment. Therefore, in a microgravity environment, it is easier to study the processes covered by gravity on the ground and the complex problems that are difficult to crack because of the coupling of gravity. To adapt to the development of the Chinese aerospace industry, the National Microgravity Laboratory of the CAS was established, which is an important measure for the development of national high-tech. During the Eleventh Five-Year Plan period, microgravity science has been closely combined with the national scientific and technological strategic objectives and critical issues of manned space flight to promote the development of high technologies such as bioengineering and new materials, as well as basic research on gravity theory and life science, and conduct preliminary research for satellite model tasks; through full demonstration, the space experiment projects with significant application value and scientific significance were selected, and extensive research work was done for the first microgravity science and space life science experiment satellite, enabling the research in this field to continue to develop steadily.

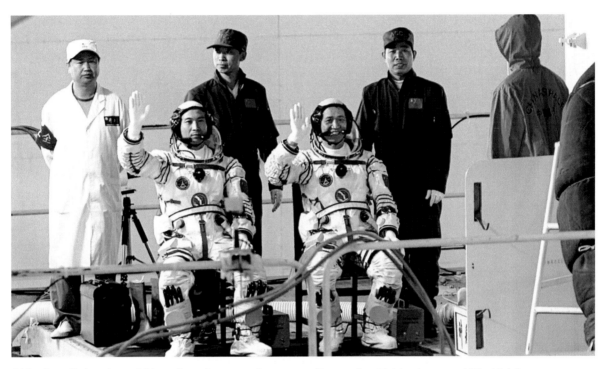

China's self-developed Shenzhou 6 manned spacecraft, carrying Fei Junlong and Nie Haisheng, was launched at Jiuquan Satellite Launch Center on October 12, 2005.

## China Manned Space Program

The Chinese space industry has developed under relatively weaker basic industries, relatively backward science and technology, special national conditions, and specific historical conditions. China has independently carried out space activities. With less investment, in a relatively shorter period, it has walked out a development path that suits its national conditions, has its own characteristics, and has made several important achievements. In many critical technical fields, such as satellite recycling, one rocket that carries multiple satellites, cryogenic fuel rocket technology, strap-on rocket technology, and geostationary orbit satellite launch and measurement and control, China stands among the world's advanced ranks. Significant achievements have been made in developing and applying remote sensing satellites and communication satellites, manned spacecraft tests, and space microgravity experiments.

The manned space program is another remarkable feat for the Chinese nation to bravely climb the peak of science and technology despite difficulties and perils. It has made China the third country in the world to independently develop manned space technology after Russia and the United States. It has raised China's status as a major space power, greatly enhanced the pride and cohesion of the Chinese nation, and is of great significance in stimulating the enthusiasm of the Chinese people to build a well-off society. Meanwhile, the manned space program has also made China an essential member of the International Space Club, providing a sustainable development impetus for the upcoming lunar and deep space exploration project.

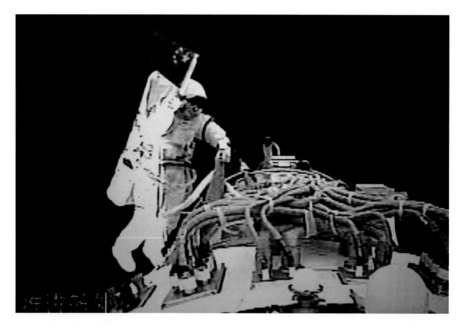

Zhai Zhigang, an astronaut of the Shenzhou 7 spacecraft, completed the spacewalk and returned to the orbital module, marking the first successful spacewalk in Chinese history.

On the afternoon of September 27, 2008, the flight crew made up of Zhai Zhigang (middle), Liu Boming (right), and Jing Haipeng (left) completed the Shenzhou 7 manned space flight mission.

## Smooth Implementation of Phase I of China's Lunar Exploration Project (CLEP)

After ten years of preparation, the CLEP was finally divided into three stages: "circling," "falling," and "returning." Chang'e-1, the first lunar exploration satellite successfully developed and launched by China, took off on October 24, 2007. After entering the lunar orbit, it performed multiple lunar explorations as planned and transmitted back a large amount of data. On March 1, 2009, it ended its mission by colliding with the moon in a controlled way.

On October 1, 2010, the Chang'e-2 satellite was successfully launched into space by a Long March 3C carrier rocket at Xichang Satellite Launch Center. The satellite carried seven kinds of detection equipment: CCD stereoscopic camera, laser altimeter, gamma-ray spectrometer, X-ray spectrometer, microwave detector, solar high-energy particle detector, and a solar wind ion detector.

The Chang'e-2 mission is a critical link between the first and second phases of the CLEP. This mission has achieved a series of engineering goals and scientific objectives. It has not only broken through a number of core technologies and key technologies and achieved a series of major scientific and technological innovations but also driven the in-depth development of China's basic science and applied technology, promoted the cross-integration of information technology and industrial technology, further formed and accumulated the management mode and experience of major scientific and technological projects with Chinese characteristics, and cultivated high-quality scientific and technological talents and management talents. This is of tremendous significance for the in-depth development of deep space explorations, the promotion of China's aerospace industry, and the construction of advanced national defense science and technology industry.

The image of the moon's surface taken by Chang'e-1.

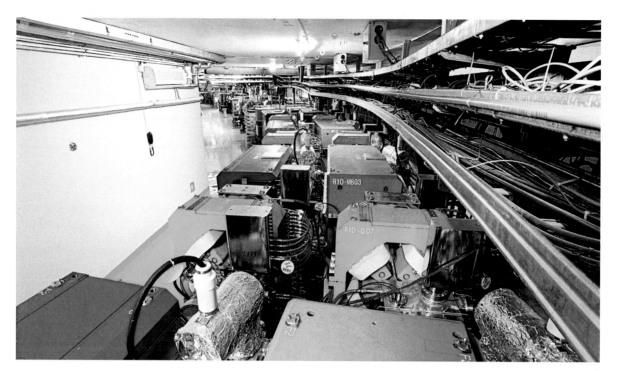

Double storage rings in the narrow tunnel—BEPC storage ring tunnel is designed according to a single storage ring. To save money and fully use the original equipment, engineers have carefully designed and installed two storage rings in the narrow tunnel and retained the existing front-end area of the synchronous radiation experiment optical beam.

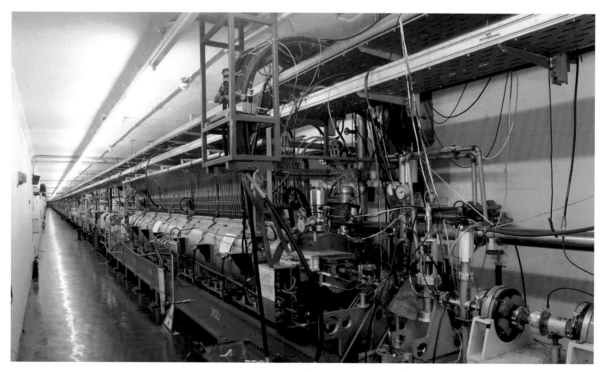

Panorama of electron linear accelerator (LINAC)—after the transformation, the positron current intensity of the 202 m long LINAC has been increased by ten times and presents excellent beam quality. As the injector of BEPC II, the injection rate of positrons into the storage ring has been raised tens of times.

BES III installation in place—scientific researchers have designed and manufactured the third-generation detector of China's high-energy physics experiment, BES III, with a total length of 11 meters, a width of 6.5 meters, a height of 9 meters, a total weight of about 800 tons, and a positioning accuracy of 2 millimeters, by using advanced detector materials and integrating the manufacturing technologies and processes of high-energy physics experiment detectors at home and abroad.

## (4) Strengthen Sharing Mechanism and Build Scientific and Technological Infrastructure and Condition Platform

### UHV AC Test Base

On February 13, 2007, the 1,000 kV single circuit test section of the ultra-high voltage (UHV) AC test base of State Grid Corporation of China was activated in Wuhan, marking the official operation of the UHV AC test base in China.

### Beijing Electron Positron Collider (BEPC) Completed Upgrading

In 2004, the Beijing Electron Positron Collider successfully finished its scheduled scientific mission. To adapt to the development of high-energy physics in the world and continue to maintain the scientific competitiveness of the BEPC, the Institute of High Energy of the CAS has decided to perform a major transformation of the BEPC with the approval of the state. As one of China's major scientific projects during the Tenth Five-Year Plan period, the BEPC Transformation (BEPC II), which cost a total of 640 million yuan, passed the national inspection and acceptance at the Institute of High-energy Physics of the CAS on the afternoon of July 17, 2009, and won the first prize of the 2016 National Prize for Progress in Science and Technology. This transformation includes injector transformation, new collider with double storage rings, new BES III and general facility transformation, etc. The high technical difficulty and complex engineering involve high-tech technologies such as low-temperature superconductivity, high-frequency microwave, ultra-high vacuum, precision machinery, magnet and power supply, beam measurement and control, particle detector, fast electronics, massive data acquisition, data-intensive computing, etc. The upgrading and transformation of large scientific research equipment such as the BEPC have become a critical "window" for learning and introducing advanced foreign science and technology, providing more opportunities and conditions for broader international cooperation and greatly expanding the channels and forms of China's international cooperation.

## Heavy Ion Research Facility in Lanzhou Cooler Storage Ring (HIRFL-CSR) Passed the National Inspection and Acceptance

On July 30, 2008, HIRFL-CSR, a major national scientific project undertaken by the Institute of Modern Physics of the CAS, passed the national inspection and acceptance. The project has completed the construction task in an all-around and high-quality way and achieved the acceptance indicators. The energy and current intensity of the main ring accelerating carbon and argon beam exceed the design indicators, lifting the cooler storage ring of China's large heavy ion accelerator to the advanced international level. This is another major achievement in China's high-tech field. HIRFL-CSR is a multi-functional, high-tech experimental device that integrates the advanced technologies of modern accelerators such as acceleration, accumulation, cooling, storage, internal target experiment, and high-sensitivity, high-resolution measurement. It has a total of 600 square meters of beam pipe with ultra-high vacuum, a total of 1,500 tons of magnet, and nearly 300 magnet power supplies. The device uses the original HIRFL system as the injector. It adopts the method of combining multi-circle injection, stripping injection, and electronic cooling to accumulate the beam in the CSRm to high current intensity and accelerate, then quickly draw it out to hit the primary target to generate a radioactive secondary beam or strip it into a highly ionized beam and inject it into the CSRe for internal target experiment or high-precision quality measurement.

## (5) Create a Favorable Environment and Strengthen Science Popularization and the Construction of Innovative Culture

### Promulgation of the *Law of the People's Republic of China on Popularization of Science and Technology*

On June 29, 2002, the 28th Meeting of the Standing Committee of the Ninth National People's Congress of the PRC adopted the *Law of the People's Republic of China on Popularization of Science and Technology* (hereinafter referred to as the *Law on Popularization of Science and Technology*). This is a milestone in the history of science popularization in China, marking that science popularization has stepped onto the track of legalization. The *Law on Popularization of Science and Technology* is an important law formulated according to China's national conditions on the basis of the decades of practice of science and technology popularization policies in China. Its promulgation is of great significance for implementing the strategy of invigorating China through science and education and sustainable development, strengthening the popularization of science and technology, improving the scientific and cultural literacy of all Chinese people, and promoting economic development and social progress. To further implement and spread the *Law on Popularization of Science and Technology*, the Education, Science, Culture, and Health Committee of the National People's Congress has specially organized the preparation of the *Interpretation of the Law on Popularization of Science and Technology*, which is a reference for all sectors of society, especially the science popularization community, to carry out work and study.

Law of the People's Republic of China On Popularization of Science and Technology

中华人民共和国
科学技术普及法

科学普及出版社
Popular Science Press

The *Law of the People's Republic of China on Popularization of Science and Technology* published by the Popular Science Press.

## Promulgation of the *Outline of the National Scheme for Scientific Literacy for All Chinese Citizens (2006–2010–2020)*

In accordance with the spirit of the 16th National Congress of the CPC, the Third, Fourth, and Fifth Plenary Sessions of the 16th Central Committee, the *Outline of the National Scheme for Scientific Literacy for All Chinese Citizens (2006–2010–2020)* (hereinafter referred to as the *Outline for Scientific Literacy*) was formulated and implemented according to the *Law of the People's Republic of China on the Popularization of Science and Technology* and the *Outline of the National Medium and Long-Term Plan for Science and Technology Development (2006–2020)*.

The *Outline for Scientific Literacy* states that "scientific literacy is an important part of citizens' literacy. Citizens' basic scientific literacy generally means understanding necessary scientific and technological knowledge, mastering basic scientific methods, establishing scientific ideas, advocating scientific spirit, and being able to apply it to solve actual problems and participate in public affairs." This is a definition of the connotation of scientific literacy based on the comprehensive analysis of the definition of scientific literacy in Chinese and foreign academic circles and from the basic national conditions. Improving Chinese citizens' scientific literacy is of great significance to enhance their ability to acquire and use scientific and technological knowledge,

With the promulgation of the Law on Popularization of Science and Technology, Xi'an City requires its primary and secondary schools to establish science and technology offices, which play a key role in promoting the popularization of scientific and technological knowledge among teenagers and improving their scientific and technological capabilities.

improve their quality of life, achieve all-round development; to improve national independent innovation capability, make China innovative, achieve comprehensive, coordinated, and sustainable economic and social development, and build a socialist harmonious society.

Our predecessors in the scientific community attached great importance to popular science education. The illustration shows Mao Yisheng, a bridge expert, explaining scientific knowledge to children.

## Science Popularization Vehicles

In 2000, the CAST began to develop and produce science popularization vehicles. By the end of 2008, 190 science popularization vehicles had been distributed nationwide, including Type I and Type II. They have covered more than 7.6 million kilometers and given over 20,000 science popularization services, benefiting more than 28 million people in total. The Type III vehicle developed in 2008 was a themed science popularization vehicle. For example, the two in the Inner Mongolia pilot tour had their own themes, "save energy and resources" and "protect the ecological environment." They carried science experience, hands-on experiments, science popularization exhibition boards, cartoons, and other science popularization resources related to their own themes. Integrating exhibition, interactive participation, and science popularization drama, they were welcomed by the general public and science popularizers and are known as mobile science museums.

A science popularization vehicle in Xinjiang.

A science popularization vehicle in Xizang.

## Science and Technology Week and Science Popularization Day

Science and Technology Week is a nationwide mass science and technology activity set up and organized by the state in 2001. It takes place in the middle of May every year. It is significant in carrying forward the scientific spirit, spreading scientific ideas, popularizing scientific knowledge, and advocating scientific methods. It plays a positive role in promoting the scientific and cultural literacy of the Chinese people and building a well-off society in an all-around way.

Since 2003, the CAST has organized associations and societies of science and technology at all levels to organize activities on science popularization day nationwide. To continue to do well in this mass and social science popularization, the CAST has decided to make the public holiday of the third week of September every year the National Science Popularization Day from 2005. The purpose was to create a positive atmosphere of "everyone further can popularize science everywhere" in the whole society by organizing and carrying out the collection of National Science Popularization Day logo and themes, stimulating all citizens to learn science, love science, and use science enthusiasm, and provide an inexhaustible source and power for the sustainable development of China's science popularization.

On October 18, 2008, The 2008 Hangzhou National Science Popularization Day Robot Exhibition was held in the Hangzhou Science and Technology Exchange. More than ten robots of various shapes and functions were exhibited, attracting many citizens to visit.

## The Newly Completed China Science and Technology Museum

On May 9, 2006, the foundation stone laying ceremony was held for the new China Science and Technology Museum, a large-scale science popularization venue invested in by the government during the Eleventh Five-Year Plan period. The new building is jointly designed by Beijing Architectural Design and Research Institute and RTKL International Co., Ltd. Its main body is a single large square. Several blocks, like building blocks, are used to engage each other, giving the whole building a huge Burr puzzle shape, reflecting the internal relationship between man and nature and science and technology. It also symbolizes that science has no absolute boundaries, and disciplines are integrated

The newly completed China Science and Technology Museum.

and mutually promoted. The new museum is located in the National Olympic Park, covering an area of 48,000 square meters and a building scale of 102,000 square meters. It is one of the related ancillary facilities of the 2008 Beijing Olympic Games and an important part of the three concepts of Green Olympics, People's Olympics, and High-tech Olympics. The construction of the new museum is another milestone in developing China's science and technology museum and a major event in China's science popularization cause. It plays an important role in strengthening China's science popularization capacity, improving citizens' scientific literacy, invigorating China through science and education, strengthening China through talents, publicizing the scientific concept of development, and building a socialist harmonious society.

# Innovation Leads the Making of a Powerful Nation of Science and Technology

VI

Since the 18th National Congress of the CPC, under the strong leadership of the Central Committee of the CPC and with the joint efforts of the Chinese science community and all other sectors of society, China's science and technology have taken giant steps and leaped significantly forward, achieving historic, overall and structural changes. Major fruits of innovation have emerged one after another. Some cutting-edge fields have begun to enter the parallel and leading stage. China's scientific and technological strength is crucial for leaping from quantitative accumulation to qualitative superiority, from point-to-point breakthrough to stronger system capability.

At the 19th Academician Conference of the CAS and the 14th Academician Conference of the CAE, General Secretary Xi Jinping stressed that to realize the great goal of building a socialist modern power and the Chinese Dream of the great rejuvenation of the Chinese nation, we must possess strong scientific and technological strength and innovation capabilities.

# Improve the Capability of Independent Innovation and Promote the Long-Term Development of Science and Technology—Implementation of the *National Development Plan for Science and Technology during the Twelfth Five-Year Plan Period*

In July 2011, the Ministry of Science and Technology of the PRC, together with relevant departments and units, completed *The National Development Plan for Science and Technology during the Twelfth Five-Year Plan Period* (after this, referred to as *The Science and Technology Plan of the Twelfth Five-Year Plan*). *The Science and Technology Plan of the Twelfth Five-Year Plan* insists on taking the realization of innovation-driven development as the fundamental mission, taking the promotion of the transformation of scientific and technological achievements into real productive forces as the main direction, taking the benefits of science and technology to the people's livelihood as the essential requirement, taking the enhancement of the long-term development capacity of science and technology as the strategic focus, and taking the deepening of reform and the expansion of opening up as a powerful driving force. It proposes that the overall aims of science and technology development during the twelfth five-year plan period are to significantly improve independent innovation capability, greatly enhance scientific and technological competitiveness and international influence, make significant breakthroughs in core and critical technologies in crucial fields, provide strong support for accelerating the transformation of the mode of economic development, and basically build a national innovation system with clear functions, reasonable structure, sound interaction, and efficient operation, lift the world ranking of the national comprehensive innovation capacity of China from the 21st to the top 18, and the contribution rate of scientific and technological progress to 55%, and make substantial progress making China innovative. At the same time, specific objectives and targets have been put forward from the aspects of R&D investment, original innovation capability, the combination of science and technology and economy, the benefits of science and technology to people's livelihood, the layout of innovation base construction, the cultivation of scientific and technological talents, and the innovation of systems and mechanisms, etc.

Tiangong-1 and the rocket combination are in place and ready for launch.

# (1) Major Fruits of Innovation

Since the Twelfth Five-Year Plan, especially since the 18th National Congress of the CPC, the Central Committee of the CPC and the State Council have attached great importance to scientific and technological innovation and made major decisions and arrangements for the in-depth implementation of the innovation-driven development strategy. China's scientific and technological innovation has therefore entered a new stage where it is catching up, running parallel, and leading at the same time, specifically, at a crucial period of leaping from quantitative accumulation to qualitative superiority, from point-to-point breakthrough to stronger system capability. As a result, its core position in overall national development has become more prominent, and its status in the global innovation landscape has been further improved, making China a major scientific and technological power with significant influence. China has made major innovative achievements in manned spaceflight and lunar exploration projects, manned deep submergence, deep drilling, supercomputing, quantum anomalous Hall effect, quantum communication, neutrino oscillation, induction of multi-functional stem cells, etc.

## Steady Implementation of the Three-Step Strategy of the Manned Space Program

Tiangong-1, the first target aircraft in China, was launched at the Jiuquan Satellite Launch Center in Gansu Province on September 29, 2011. The launch vehicle, with a total length of 10.4 meters and a maximum diameter of 3.35 meters, comprises an experimental module and a resource module. Its launch marks the second step and the second stage of China's three-step space strategy.

On November 3, 2011, the successful rendezvous and docking of the Tiangong-1 and Shenzhou 8 spacecraft was an important step in China's breakthrough and mastery of space rendezvous and docking technology, marking that China has become independently capable of space rendezvous and docking and building its own space laboratory, that is, a short-term unattended space station. On June 18, 2012, the manned rendezvous and docking mission of the Tiangong-1 and Shenzhou 9 spacecraft succeeded, another major breakthrough in China's space rendezvous and docking technology, marking the decisive and important progress made in the second strategic goal of China's manned space program. On June 11, 2013, the Shenzhou 10 spacecraft successfully took off and docked with Tiangong-1 on June 13. On June 20, astronaut Wang Yaping explained and demonstrated the characteristics of object motion and liquid surface tension in a weightless environment, setting a precedent for an applied flight of China's manned space programs.

On June 20, 2013, the astronauts of Shenzhou 10 carried out basic physics experiments on Tiangong-1, giving a class to teenagers from space. More than 60 million teachers and students from more than 80,000 middle schools nationwide tuned in to watch the live stream.

## The Lunar Exploration Project in a New Stage

The Chang'e-3 probe was launched into space by Long March 3B launch vehicle at the Xichang Satellite Launch Center of China on December 2, 2013. On December 14, it made a soft landing on the moon and started to carry out scientific exploration missions one after another.

As the primary mission of the second phase of the lunar exploration project, Chang'e-3 was going to complete three major engineering objectives and three types of scientific exploration tasks. The three primary engineering objectives include the following:

1. Break through the key technologies such as lunar soft landing, lunar surface patrol and survey, deep space TT&C (Telemetry, Tracking, and Command) communication and teleoperation, the launch of deep space exploration launch vehicles, to improve the level of space technology.

2.  Develop lunar soft landing probe and patrol probe, establish ground deep space station, obtain functional modules, including launch vehicle, lunar probe, launch site, deep space TT&C station, ground application, etc., to have the basic capability of a lunar soft landing and lunar exploration.

3.  Establish the basic system of lunar exploration and space engineering, and form scientific and effective engineering methods for the implementation of major projects. The three types of scientific exploration tasks include lunar surface topography and geological structure survey, investigation of monthly surface material composition and available resources, and earth's plasma layer detection and lunar-based optical astro-observation.

On the evening of December 15, 2013, the Chang'e-3 lander and rover, carrying out scientific exploration on the moon, conducted a mutual imaging experiment and took photos of one another. The illustration is the photo of the Yutu lunar rover shown on the big screen of the Beijing Flight Control Center, which was taken by the camera on the Chang'e-3 lander.

The effective mutual imaging between the Chang'e-3 lander and rover marks the successful completion of the second phase of China's lunar exploration project. The Chang'e-3 mission has achieved the first soft landing and patrol and survey of Chinese spacecraft on extraterrestrial objects, marking the overall triumph of the second step of the strategic goals of "orbiting, landing and returning" of China's lunar exploration project. This was an important milestone in the development of the Chinese space industry. With the triumph of the Chang'e-3 mission as a symbol, China's lunar exploration project has entered a new stage of unmanned automatic sampling and return.

The return vehicle for the re-entry and return flight test was launched at Xichang Satellite Launch Center and entered the Earth-Moon transfer orbit on October 24, 2014. It effectively implemented two orbit corrections, arrived at the lunar gravitational sphere on the 27th, and began the veered flight near the moon. On the evening of the 28th, it completed the veered flight near the moon and entered the Moon-Earth transfer orbit. On the 30th, it successfully implemented once more orbit correction and returned to earth. The success of the first re-entry and return flight test marks that China has comprehensively broken through and mastered the critical technology of re-entry and return of spacecraft at a speed close to the second universe speed (11.2 km/s). This has laid a solid foundation for the full completion of the three-step strategic goal of "orbiting, landing, and returning" of the lunar exploration project and is of great significance to the sustainable development of China's lunar and deep space exploration and even its aerospace undertakings.

On November 1, 2014, after a journey of several days between the earth and the moon, the return vehicle for the re-entry and return flight test of the third phase of China's lunar exploration project landed safely in the predetermined area of Siziwang Banner, Inner Mongolia.

## Smooth Takeoff of New Launch Vehicles

On September 20, 2015, the Long March 6 launch vehicle took off at Taiyuan Satellite Launch Center. Long March 6 is a three-stage liquid launch vehicle. Its power system adopts a liquid oxygen kerosene engine, which is non-toxic and pollution-free and takes a short launch preparation time. It is mainly used to meet the launch requirements of microsatellites. The R&D of this launch vehicle was completed by the Shanghai Academy of Aerospace Technology, affiliated with China Aerospace Science and Technology Corporation. Filling the gap between non-toxic and pollution-free launch vehicles in China, it is of great significance to improve the range of types of Chinese launch vehicles, improve the safety and environmental protection of rocket launch, and enhance the ability to enter space.

The Long March 6 launch vehicle took off from Taiyuan Satellite Launch Center, successfully sending 20 micro-satellites into space. The success of this launch mission not only marks the addition of a new member to the Long March series but also sets a new record for China's space launch of one rocket with multiple satellites.

## Leading in Manned and Unmanned Deep Submergence

Deep-sea research is a global hotspot. The deep sea below 1,000 meters holds abundant resources, including metallic ores, deep-sea gas, and biological genes, which are essential strategic resources for future human development. Meanwhile, there is great scientific research value in the deep sea in many fields, such as earth science, life science, environmental science, etc. During the Twelfth Five-Year Plan period, China has made outstanding achievements in deep-sea diving.

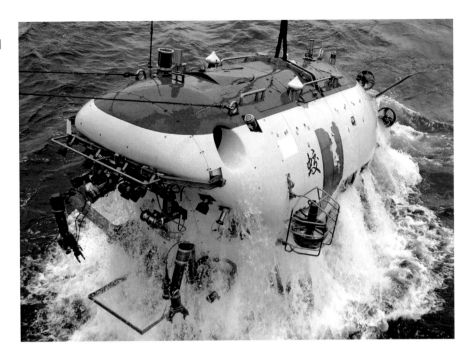

The Jiaolong manned submersible conducted the fourth 7,000-meter-level sea trial dive test.

## 1) Jiaolong dived 7,062 meters

On June 24, 2012, The Jiaolong manned submersible set a new record of 7,062 meters of Chinese manned deep diving in the Mariana Trench test area in the western Pacific Ocean, which is the largest diving depth of the same type of manned submersible in the world, indicating that China is able to perform scientific research and resource exploration in 99.8% of the world's deep seas. At the beginning of the Jiaolong manned submersible project in 2002, China's manned deep diving technology could only dive to a few hundred meters. The leap from a few hundred meters to 7,000 meters was a change of distance and a journey of several generations of Chinese manned deep-diving researchers and deep-sea scientists.

## 2) The success of the sea trial of Haima

The Haima ROV is the main scientific research achievement of the 4,500-meter-level deep sea operation system project in the marine technology field of Project 863. From February to April 2014, it successfully dived to 4,502 meters deep in the central basin of the South China Sea. It completed a number of deep-sea geological exploration tasks, such as simulation deployment of submarine observation network extension cable, sediment sampling, heat flow probe detection, OBS deployment, seabed self-photography, and marker deployment, among other deep-sea geological exploration missions. It also accomplished the joint operation with the underwater lifting device, passing the on-site assessment and sea trial inspection and acceptance that involved 91 technical indicators. The success of the sea trial of the Haima submersible marks that Chinese marine technicians have comprehensively broken through and mastered the core technologies related to the deep-sea ROV, ending the unfavorable situation in which other countries have long controlled the technology and equipment of the deep-sea ROV.

*3) 10,000-meter deep diving of Tansuo No. 1*

The Tansuo No. 1 scientific research ship conducted scientific research in the Challenger Abyss of Mariana Trench from June 22 to August 12, 2016. The expedition lasted 52 days. In this scientific investigation, the research team carried out 84 scientific research tasks in the Mariana Trench area and obtained a wealth of precious samples and data in different depth sections, using a series of high-tech equipment such as the 10,000-meter autonomous remote control submersible Haidou, the abyssal landers Tianya and Haijiao, the 10,000-meter in-situ test system Yuanwei Shiyan, the 9,000-meter deep-sea seabed seismometer, the 7,000-meter deep-sea glider and so on, which were all independently developed by China.

Tansuo No. 1 sailed away.

## Chinese Supercomputers Dominate the World

On June 20, 2016, in the newest Top 500, a respected list of the world's most powerful computers, Sunway TaihuLight, made with Chinese chips, replaced Tianhe-2 (Milky Way-2) and topped the list. For the first time in history, the total number of Chinese supercomputers on the list exceeded that of the United States, ranking first. The floating point arithmetic speed of Sunway TaihuLight is 93 quadrillion times per second, which is not only nearly twice as faster as the second-place Milky Way-2, but its efficiency is also three times higher than the latter. More importantly, unlike Milky Way-2, which uses Intel chips, Sunway TaihuLight uses chips with Chinese independent intellectual property rights.

Sunway TaihuLight was developed by the National Research Center of Parallel Computer Engineering & Technology and installed in the National Supercomputing Wuxi Center. Previously, the Milky Way-2, developed by the National University of Defense Technology, had been ranked No.1 among the top 500 six consecutive times.

## Discovery of New Neutrino Oscillations

On March 8, 2012, Wang Yifang, spokesman of Daya Bay Neutrino Experiment International Cooperation Group, announced in Beijing that the Daya Bay neutrino experiment discovered a new neutrino oscillation and measured its oscillation probability. This significant achievement, being a new understanding of the basic laws of the physical world, plays a decisive role in the future development direction of neutrino physics and will help to solve the "mystery of the disappearance of antimatter" in the universe.

Xue Qikun (1963–), a material physicist and academician of the CAS, was rated as one of the most influential "ten figures of scientific and technological innovation" in 2016. The illustration shows Xue working in the laboratory of Tsinghua University.

## The Quantum Anomalous Hall Effect Was Observed for the First Time

In 2013, headed by Xue Qikun of Tsinghua University, the experimental team that consisted of the Department of Physics of Tsinghua University and the Institute of Physics, CAS observed the anomalous quantum Hall effect for the first time. Their research findings were published online in *Science* magazine on March 14, 2013.

## Breeding of Cloned Mice via Induced Pluripotent Stem Cells

Chinese scientists Zhou Qi, Zeng Fanyi, and others injected four genes into mouse fibroblasts using viruses as vectors, transformed them into induced pluripotent stem cells (iPS cells), and then bred a mouse embryo on this basis, and implanted it into the body of the experimental surrogate mouse.

Twenty days later, a small black mouse was born, and genetic tests confirmed that the mouse's offspring provided fibroblasts. This mouse, which was successfully bred for the first time, was named Tiny. According to the press release by the British journal *Nature*, Chinese scientists have cloned a complete live experimental mouse with iPS cells for the first time, thus verifying for the first time that iPS cells are as versatile as embryonic stem cells.

## Fast Neutron Experimental Reactor Combined to the Grid

On July 21, 2011, the first Chinese experimental fast reactor with nuclear fission caused by fast neutrons was successfully combined with the grid for power generation. The full realization of the objectives of this significant project of Project 863 marks a major breakthrough in fast reactor technology included in the frontier technology of the *Outline of the National Medium and Long-Term Plan for Science and Technology Development (2006–2020)*, as well as an important step in China's occupation of the commanding heights of nuclear energy technology and establishment of an advanced nuclear energy system for sustainable development. A fast neutron reactor is the main reactor type of the world's fourth-generation advanced nuclear power system. China Experimental Fast Reactor is the first step in China's development of a fast neutron breeder reactor. This reactor adopts an advanced pool structure. With nuclear thermal power of 65 MW and experimental generated output of 20 MW, it is one of the few high-power experimental fast reactors with power generation functions worldwide. Its main system settings and parameter selection are the same as those of large fast reactor power plants. The experimental fast reactor fully uses its inherent safety and adopts various passive safety techniques. Its security has met the requirements of the fourth-generation nuclear power system.

## Discovery of Protein Kinase GRK5

After more than three years of research, the research team headed by Professor Ma Lan from the Institute of Brain Science, Fudan University, discovered that GRK5, a protein kinase widely existing in the body, plays a crucial role in neural development and plasticity. If the mouse lacks GRK5, it will suffer abnormal neuronal morphological development and show obvious cognitive deficits, memory impairment, and learning ability reduction. At the same time, this effect of GRK5 has nothing to do with its protein kinase function.

## Discovery of Iron-Based High-Temperature Superconductors

At the 2013 State Science and Technology Awarding Meeting of the PRC, the research team, represented by Zhao Zhongxian, Chen Xianhui, Wang Nanlin, Wen Haihu, and Fang Zhong of the Institute of Physics, CAS and the University of Science and Technology of China, won the first prize of the 2013 State Natural Science Award for their outstanding contributions to the "Discovery of Iron-based High-temperature Superconductors above 40 K and Research on Some Basic Physical Properties."

Superconductivity, one of the greatest scientific discoveries in the 20th century, refers to the phenomenon that some materials suddenly show no resistance when cooled to a specific critical temperature. Materials with this property are called superconductors. According to traditional theoretical calculations, physicist William McMillan deduced that the transformation temperature of superconductors should not exceed 40 K (about −233 °C), which is also known as the McMillan limit temperature. The Chinese research team first found iron-based superconductors with transformation temperatures above 40 K, breaking through the McMillan limit temperature, and identified iron-based superconductors as a new type of high-temperature superconductors. Then they discovered a series of iron-based superconductors with transformation temperatures above 50 K and set a world record of 55 K. Academician Zhao Zhongxian, who has been studying superconducting since 1964, has played a crucial role in this team.

Zhao Zhongxian: the leader of superconductivity studies in China.

## Discovery of Secretase in Alzheimer's Disease

On August 18, 2015, the research team led by Professor Shi Yigong of the School of Life of Tsinghua University published an online essay entitled "An Atomic Structure of Human γ- Secretase" in *Nature*. It reports the electron microscopic structure of human γ-secretase with a resolution of 3.4 Angstrom and studied, based on structural analysis, the function of pathogenic mutants of the γ-secretase, thus providing an essential basis for understanding the working mechanism of γ-secretase and the pathogenesis of Alzheimer's disease.

Shi Yigong is an academician of the CAS and a structural biologist, mainly engaged in the molecular mechanism of apoptosis, important membrane proteins, and the structure and function of intracellular biomacromolecule machinery.

## Discovery of Weyl Fermions

In 2015, the team led by researcher Fang Zhong from the Institute of Physics, CAS, discovered Weyl fermions in the experiment for the first time. It is the first time to confirm the existence of the Weyl fermion state in condensed matter since it was proposed in 1929, which has great scientific value. Scientists believe that this discovery is of great significance for breakthroughs in disruptive technologies such as topology electronics and quantum computers. Chinese scientists independently completed the entire discovery, from theoretical prediction to experimental observation. The Weyl Fermions Study was selected into the Top Ten Breakthroughs in 2015 of *Physical World*.

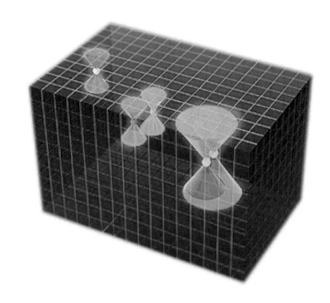

The figure shows three types of fermions experimentally discovered in solid materials: quadruple-simplified Dirac fermions (left), two-simplified outer fermions (center), and triple-simplified novel fermions (right).

## Realization of the Teleportation of Quantum System with Multiple Degrees of Freedom

The research team composed of Pan Jianwei of the University of Science and Technology of China and his colleagues Lu Chaoyang, Liu Naile, and others has succeeded in realizing the teleportation of a quantum system with multiple degrees of freedom for the first time in the world. On February 26, 2015, *Nature* published this latest research finding in the form on the cover. This is another important breakthrough scientists have made in the field of quantum information experimental research after 18 years of efforts since the first international realization of quantum teleportation with a single degree of freedom in 1997, thus laying a solid foundation for the development of scalable quantum computing and quantum network technology.

## Biomedical Development Improves Public Well-Being

During the Twelfth Five-Year Plan period, biomedical innovation has provided a strong guarantee for improving public well-being: the world's first bioengineered cornea Ai Xin Tong was released; the world's first gene mutant Ebola vaccine was clinically tested overseas; inactivated poliomyelitis vaccine against hand-foot-mouth disease was effectively developed; new anti-tumor drugs such as Appatinib and Chidamide have been marketed, playing an important role in alleviating the difficulty and high cost of seeking for medical help.

## Generation of Antimatter—Ultrafast Positron Source

On March 10, 2016, the Key Laboratory of Intense Field Laser Physics of Shanghai Institute of Optics and Fine Mechanics, CAS announced that it had successfully produced antimatter—ultrafast positron source by using ultra-intense and ultra-short lasers, which is also the first time that Chinese scientists have successfully created antimatter by using lasers. This discovery will have significant applications in the fields of nondestructive detection of materials, laser-driven positron collider, cancer diagnosis, etc.

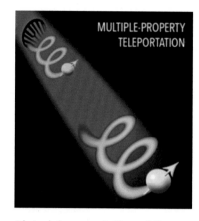

Pictorial presentation of the quantum invisible state transfer process for single-photon spin and orbital angular momentum.

## (2) Major Breakthroughs in Major Equipment and Strategic Products

During the Twelfth Five-Year Plan period, major equipment and strategic products such as high-speed railway, hydropower equipment, ultra-high voltage power transmission and transformation, hybrid rice, the fourth generation mobile communication (4G), earth observation satellite, Beidou navigation system, and electric vehicles in China have made major breakthroughs. Some products and technologies have begun to go global.

## Rapid Development of High-Speed Railway

Since the opening of the Beijing–Tianjin Intercity Railway, the first 350 km/h high-speed railway in China, on August 1, 2008, the scale of the Chinese high-speed railway has expanded explosively. As per the national medium and long-term railway network planning and the railway planning in the Eleventh Five-Year Plan and Twelfth Five-Year Plan, the construction of a high-speed railway with the "four verticals and four horizontals" fast passenger transport network as the main framework has been comprehensively accelerated. A number of high-speed railways with a design speed of 350 km/h and advanced international standards, including Beijing–Tianjin, Shanghai–Nanjing, Beijing–Shanghai, Beijing–Guangzhou, Harbin-Dalian railways, have been built, forming a relatively complete high-speed railway technology system. Through the introduction, understanding, absorption, and re-innovation, China has systematically mastered the manufacturing technology of bullet train with a speed of 200–250 km/h, successfully built a technical platform for bullet train with a speed of 350 km/h, and developed and produced a new generation of high-speed trains CRH380.

Harbin–Dalian High-Speed Railway is the first alpine high-speed railway in China and the world to be put into operation.

## Accelerate the Construction of Hydropower Equipment

The critical tasks of hydropower science and technology in the Twelfth Five-Year Plan include: first, building a Chinese hydropower science and technology innovation system to enhance national hydropower science and technology innovation capability; second, based on independent innovation, increasing the investment in strategic science and technology, solving the technical bottleneck of "safe and efficient" development and utilization of hydropower resources through major technical research and major equipment R&D, driving the leapfrog development of productivity, and improving the overall competitiveness of the industry; third, organizing the research and project demonstration of ecological restoration technology, scientifically balancing the relationship between hydropower development and ecological environment, and developing and building eco-friendly hydropower projects.

On July 10, 2014, the last unit of Xiangjiaba Hydropower Station, China's third-largest hydropower station, was officially put into operation. So far, Xiluodu and Xiangjiaba hydropower stations in the lower reaches of the Jinsha River have started working, with an installed capacity of 20.26 million kilowatts, close to that of the Three Gorges Hydropower Station. As the national critical power supply project of power transmission from west to east, the two power stations continuously send clean energy to East China, Central China, and Southern Power Grids and cater to the power demand of Sichuan and Yunnan in the dry seasons. Their designed annual average power generation is 88 billion kilowatt hours, which saves more than 33 million tons of standard coal, thus reducing the emission of carbon dioxide by more than 75 million tons, sulfur dioxide by about 900,000 tons, and nitrogen oxides by about 390,000 tons.

## Comprehensively Master Ultra-High Voltage Power Transmission and Transformation Technology

On January 18, 2013, at the 2012 State Science and Technology Awarding Meeting of the PRC held in Beijing, nearly 50,000 people from more than 100 units, including the State Grid Corporation of China, participated in the R&D, and construction of the "Key Technologies, Complete Sets of Equipment and Engineering Applications of Ultra-high Voltage (UHV) Alternating Current Transmission" project, which won the special prize of the National Scientific and Technological Progress Award. The completion of this project marks that China has fully mastered the UHV core technology. During the project implementation, a complete UHV standard system and specifications were developed, realizing the two goals of "Made in China" and "Led by China" and enhancing the international competitiveness of the Chinese electric power equipment manufacturing industry.

## Major Breakthrough in the Third Phase of Super Rice

Rice is one of the most important food crops in China. It is a crucial choice to solve the problem of food security, relying on scientific and technological progress to continue to increase rice yield per unit area substantially. In 1996, the Ministry of Agriculture launched the major science program

Research on China's Super Rice Breeding and Cultivation Technology System. In 2005, Project 863 initiated the third phase of super hybrid rice research, aimed at breeding new super hybrid rice varieties with a unit yield of 13.5 t/hm² by 2015. On September 18, 2011, after the on-site yield inspection and acceptance by the expert group of the Ministry of Agriculture, the average unit yield of Y Liangyou No. 2 planted in Yangguao Township, Longhui County, Hunan Province, reached 13.9 t/hm², marking a significant breakthrough in the third phase of super hybrid rice breeding in China four years ahead of plan. It was rated as a top ten scientific and technological progress in China in 2011.

Yuan Longping: Fighting hunger with hybrid rice.

## Earth Observing Satellite

China has built its civil aerospace infrastructure consisting of ZY series earth resource satellites, FY series meteorological satellites, HY series ocean monitoring satellites, HJ series environment and disaster monitoring small satellites, etc.

### 1) ZY series satellites

On January 9, 2012, the ZY-3 satellite was successfully launched by the Long March 4B launch vehicle at Taiyuan Satellite Launch Center. On January 11, it effectively transmitted the first group of high-precision stereo images and high-resolution multispectral images, covering Heilongjiang, Jilin, Liaoning, Shandong, Jiangsu, Zhejiang, Fujian, among other regions, with a total area of about 210,000 square kilometers. On April 20, 2012, the satellite in-orbit test was completed. The ZY-3 satellite is China's first independent civil high-resolution stereo mapping satellite. Through stereo observation, it can measure and produce 1:50,000 scale topographic maps and render services for land resources, agriculture, and forestry, among other fields, filling the gap in the field of stereo mapping in China, thus a milestone.

### 2) China High-Resolution Earth Observing System (CHEOS)

The Gaofen 1 satellite was successfully launched by the Long March 2D launch vehicle at Jiuquan Satellite Launch Center on April 26, 2013. Being a high-resolution earth observing satellite, it is equipped with two 2-meter resolution panchromatic, 8-meter resolution multispectral cameras and four 16-meter resolution multispectral broadband cameras. The Gaofen satellite system has broken through many key technologies, integrating spatial resolution, multi-spectrum, and wide coverage.

On August 19, 2014, the Gaofen 2 satellite was successfully launched by Long March 4B launch vehicle at Taiyuan Satellite Launch Center and entered its scheduled orbit. The spatial resolution of the Gaofen 2 satellite is higher than one meter, while it has the characteristics of high radiation accuracy, high positioning accuracy, and fast attitude maneuver capability. This marks that China's remote sensing satellites have entered the sub-meter "Gaofen era."

On August 10, 2016, the Gaofen 3, China's first one-meter resolution C-band multipolar SAR (synthetic aperture radar) satellite, made it to space, officially commencing its mission. It is the first multipolar SAR satellite in China. Able to perform remote sensing imaging with a maximum resolution of one meter, it is the highest resolution C-band multipolar SAR satellite worldwide. The launch and application of the Gaofen 3 satellite have brought the construction of China's Gaofen system from visible light, thermal infrared ray, and far infrared ray to microwave radiation area, thus ushering in a new era of satellite microwave remote sensing applications.

On December 29, 2015, China successfully launched the Gaofen 4 satellite via a Long March 3B launch vehicle at Xichang Satellite Launch Center. This is China's first geosynchronous orbit high-resolution remote sensing satellite. It operates in the geosynchronous orbit above the equator. Taking advantage of synchronization with the earth and relatively static to the planet, it can "stare" at the target area for a long time, obtain dynamic change process data, and perform near real-time emergency tasks such as forest fire monitoring. As an important satellite of the CHEOS national major science and technology project, the development and launch of Gaofen 4 significantly improve the overall design level of China's remote sensing satellites and the technical level of high-performance remote sensing optical payloads, opening up a new field of China's high-orbit and high-resolution earth observation technology, and greatly enhance China's space-based remote sensing observation capability.

### 3) FY series satellites

China began to develop meteorological satellites in 1977. It launched three first-generation polar-orbiting meteorological satellites in 1988, 1990, and 1999, namely FY-1A, B, and C. In 1997 and 2000, two geostationary FY-2 meteorological satellites were launched, forming the Chinese meteorological satellite operational monitoring system, making China the third country to have two kinds of meteorological satellites in orbit in the world after the United States and Russia. On May 27, 2008, China successfully launched a new generation of polar-orbiting meteorological satellite FY 3A, which has global, all-weather, three-dimensional, quantitative, and multispectral remote sensing monitoring capabilities. It has achieved four major technological breakthroughs in China's meteorological satellite, from single remote sensing imaging to comprehensive detection of the earth's environment, from optical remote sensing to microwave remote sensing, from kilometer-level resolution to hundred-meter level resolution, and from domestic reception to polar reception. On December 11, 2016, China successfully launched the FY-4 satellite (satellite 01) with Long March 3B launch vehicle at Xichang Satellite Launch Center. This not only means that China's future weather monitoring, forecasting, and early warning will be more accurate but also that China has reached the world's advanced level in high-end meteorological satellites.

### 4) HY series satellite

At 6:57 on August 16, 2011, a Long March 4B launch vehicle carrying the HY-2 satellite took off from the Taiyuan Satellite Launch Center. HY-2 is China's first marine dynamic environment monitoring satellite. It mainly monitors and investigates the marine environment, an important monitoring tool for disaster prevention and reduction. It can directly serve as the early warning of disastrous sea conditions and national economic construction and provide satellite remote sensing information for marine studies, marine environment forecasting, and global climate change research.

As of the end of 2016, HY-1 A/B and HY-2 satellites have been launched, enabling China to observe the dynamic marine environment and widely carry out marine living resources surveys, marine environment monitoring, etc.

### 5) HJ series satellites

The environment and disaster monitoring and forecasting small satellite constellation is a remote sensing satellite constellation program that China proposed to adapt to the new forms and requirements of environmental monitoring, disaster prevention, and mitigation. As required by disaster and environmental protection operations, the environment and disaster monitoring and forecasting small satellite constellation are composed of optical satellites and SAR satellites with medium to high spatial resolution, high temporal resolution, high spectral resolution, and wide observation width, which can comprehensively use visible light, infrared ray and microwave remote sensing, among other observation means. They are developed to meet the demand of disaster and environmental monitoring and prediction for time, space, spectral resolution, and all-weather all-day observation.

The "4+3" environment and disaster monitoring and prediction small satellite constellation has the disaster monitoring capability of a medium resolution, wide coverage, and high revisit, providing decision-making support for pre-disaster risk early warning, disaster emergency monitoring and assessment, and post-disaster relief, recovery and reconstruction.

Sun Jiadong with Xichang Satellite Launch Center. Sun Jiadong (1929–) is one of the pioneers of China's man-made satellite technology and deep space exploration technology and an academician of the CAS.

## The Beidou-2 Satellite Navigation System

At the 2017 State Science and Technology Awarding Meeting of the PRC, the Beidou-2 satellite project won the National Scientific and Technological Progress Award special prize. It is a major special national science and technology project and a critical step in China's "three-step" development strategy for building the Beidou satellite navigation system. It aimed to build the Beidou-2 satellite navigation system that covers the Asia Pacific region, meet the urgent needs of China's economic and social development and the construction of the national defense and armed forces, and safeguard national security and strategic interests. The project was set up in August 2004. It took eight years to complete the development and construction. Consequently, the Beidou-2 satellite navigation system, which consists of 16 networking satellites and 32 ground stations that operate in a coordinated network, came into being, thus starting to render navigation, positioning, time service, and short message communication services to the Asia Pacific region in December 2012. Its system performance in the service area is as good as that of similar foreign systems, reaching the advanced international level in the same period.

## Offshore Oil Drilling Platform 981 in Operation

The construction of offshore oil deep water semi-submersible drilling platform 981 was started on April 28, 2008. It is the sixth generation's first deep water semi-submersible drilling platform, independently designed and built in China. Its completion marks China has become capable of independent R&D of ocean engineering equipment and internationally competitive. On May 9, 2012, Offshore Oil Drilling Platform 981 started operating in the South China Sea, marking a substantial step in the deep water strategy of China's offshore oil industry.

Offshore Oil Drilling Platform 981.

## The First Superconducting Substation in Operation

On April 20, 2011, the world's first superconducting substation developed by China was officially put into operation in Baiyin City, Gansu Province. At present, the world's only distribution-level fully superconducting substation has created a number of world's and China's firsts, marking a significant breakthrough in China's superconducting power technology. Compared with traditional substations, superconducting substations play an important and irreplaceable role in greatly improving reliability, safety, and quality of power supply and transmission capacity while reducing transmission loss.

The operating voltage level of this substation is 10.5 kilowatts. It integrates a variety of new superconducting power devices, such as superconducting energy storage systems, superconducting current limiter, superconducting transformers, and three-phase alternating current high-temperature superconducting cable, thus able to greatly improve the safety of the power grid and the quality of power supply, effectively reduce the system loss and the floor area.

## Successful Development of Laser Rapid Manufacturing Equipment

In 2011, Shi Yusheng's Huazhong University of Science and Technology research team successfully developed rapid laser sintering manufacturing equipment with a shaping space of 1.2 meters × 1.2 meters based on a powder bed. The equipment has been effectively applied to a variety of high-end fields, such as aerospace missions and automobile engines, playing an important role in the innovative R&D of high-end technologies for Chinese enterprises. Since the biggest advantage of rapid manufacturing technology is that there is almost no constraint on the geometric shape of the components, it is expected to revolutionize high-end manufacturing fields with complex structures, such as aerospace missions, weapons, automobiles, and other power equipment.

## Biggest Rotary Radio Telescope in Asia Completed

On October 28, 2012, the largest omnidirectional rotating radio telescope in Asia was officially completed at the Shanghai Observatory. It ranks first in Asia and fourth in the world regarding comprehensive performance. It can observe objects further than 10 billion light years away. It has participated in China's lunar exploration project and various deep space exploration projects.

## (3) Scientific and Technological System Reform

During the Twelfth Five-Year Plan period, the central government systematically promoted the reform of the science and technology system in terms of resource allocation, program management reform, a transformation of scientific and technological achievements, and talent evaluation, and has made four breakthroughs.

First, significant changes have taken place in the allocation of scientific and technological resources across society, and new economic drivers continued to emerge.

Second, important breakthroughs have been made in the reform of science and technology management, and the central government grants have been allocated more to basic research, strategic frontier, social welfare, and major projects.

Third, institutional barriers have been demolished to facilitate the transformation of scientific and technological achievements.

Fourth, the talent development environment has been further optimized, the academician system reform was conducted in an orderly manner, talent programs such as the Thousand Talents Program and the Ten Thousand Talents Program have vigorously promoted the introduction and training of high-end talents, significant progress has been made in making scientific researchers younger, and the talent use, training, and incentive mechanism have been constantly perfected.

## (4) Deepened International Scientific and Technological Cooperation

China's scientific and technological innovation strength has improved significantly during the Twelfth Five-Year Plan period. International innovation cooperation has also developed rapidly, which has played a crucial role in supporting scientific and technological innovation. China's global scientific and technological cooperation pattern is also undergoing a historic transformation. The relationship between international cooperation and scientific and technological innovation and development is changing from "passively following and assisting" to "actively laying out, supporting, and leading."

According to the *Report on the Status of China's International Science and Technology Cooperation* released by the National Center for Science and Technology Evaluation, from 2006 to 2015, the centrality of China's scientific research cooperation rose from tenth place to seventh in the world, and its scale rose from the sixth to the fourth. China is promoting the recovery and growth of the world economy with active scientific and technological innovation cooperation and making the voice of China heard.

Simultaneously, China has launched a series of science and technology partnership programs through bilateral channels to share the experience and achievements of science and technology development with developing countries. From 2001 to 2015, China trained nearly 10,000 scientific and technological personnel from more than 120 developing countries. At a series of UN summits in September 2015, President Xi Jinping announced a series of major initiatives to promote global development, which will positively contribute to promoting scientific and technological cooperation among countries and effectively implementing the development agenda.

## (5) Continuous Optimization of Innovation and Entrepreneurship Environment

During the Twelfth Five-Year Plan period, the innovation and entrepreneurship environment of the whole society has been continuously optimized, with the national innovation demonstration zone and high-tech industry development zone as important carriers of innovation and entrepreneurship. The revision and implementation of the *Law of the PRC on Promoting the Transformation of Scientific and Technological Achievements* have opened the channel for the integration of science and technology with the economy, promoted mass entrepreneurship and innovation, encouraged R&D institutions, colleges and universities, enterprises and other innovation entities and scientific and technological personnel to transfer and transform scientific and technological achievements, propelled the upgrading of economic quality and efficiency, and achieved remarkable results in the implementation of policies such as the addition and deduction of enterprise R&D expenses. Consequently, the integration of science and technology with finance has become greater, the Chinese citizens' scientific literacy has been steadily improved, and the entire society's innovation consciousness and vitality have been significantly more robust.

### Zhongguancun Science Park

Haidian District of Beijing is one of the first batches of national mass entrepreneurship and innovation demonstration bases. The high-tech industry accounts for more than 60% of the GDP of the whole district. As the "experimental field" of scientific and technological innovation reform, from

Zhongguancun Electronics Street to New Technology Industry Development Pilot Zone, from the first national high-tech zone to the first national innovation demonstration zone, Zhongguancun closely follows the wave of technological revolution, breaks through the shackles of the system and mechanism, and walks out of a path of pioneering spirit and commitment to innovation.

## Wuhan East Lake National Independent Innovation Demonstration Zone

The Wuhan East Lake Demonstration Zone, known as the Optics Valley of China, is the second national independent innovation demonstration zone in China after Zhongguancun. There are nearly 100 universities and research institutes, more than 30 national key laboratories, and the Wuhan National Laboratory for Optoelectronics, representing the highest level of optoelectronic technology in Asia. Adhering to the road of independent innovation, the East Lake Demonstration Zone has formed a high-tech industrial cluster with an optoelectronic information industry as the core includes biomedicine, energy and environmental protection, and geospatial information by cultivating high-tech enterprises and developing emerging industries. At present, it houses more than 20,000 high-tech enterprises. In 2012, they generated a total income of 500.6 billion yuan, and their leading economic indicators maintained an average annual growth of 30%.

# Led by Scientific and Technological Innovation, Open New Development Realm— Implementation of *Thirteenth Five-Year Plan on Scientific and Technological Innovation*

On the morning of May 30, 2016, the National Conference on Science and Technology Innovation, the 18th Academician Conference of the CAS, the 13th Academician Conference of the CAE, and the 9th National Congress of the CAST were grandly held in the Great Hall of the People. At the meeting, General Secretary Xi Jinping delivered an important speech emphasizing that science and technology is a powerful tool that makes a country powerful, makes an enterprise depends victorious, and improves people's lives. If China is to be strong and the Chinese people are to live a better life, China must have strong science and technology. The new times, situations, and tasks require China to have new ideas, designs, and strategies for scientific and technological innovation. To realize the "Two Centenary Goals" and the Chinese Dream of the great rejuvenation of the Chinese nation, China must adhere to the path of independent innovation with Chinese characteristics, accelerate scientific and technological innovation in all fields, and seize the opportunities of global scientific and technological competition. This is the starting point for making China a  science and technology power in the world.

The *Thirteenth Five-Year Plan for National Scientific and Technological Innovation* (from now on referred to as the *Thirteenth Five-Year Plan for Science and Technology*), formulated based on the *Outline of the Thirteenth Five-Year Plan for National Economic and Social Development of the People's Republic of China*, the *Outline of the National Innovation-driven Development Strategy*, and the *Outline of the National Medium-and Long-Term Plan for Science and Technology Development (2006–2020)* is a key special plan in the field of scientific and technological innovation and an action guide for China to become an innovative country.

The Thirteenth Five-Year Plan period is a decisive stage for building a moderately prosperous society in all respects and entering the ranks of innovative countries. It is also a critical period for deepening the implementation of the innovation-driven development strategy and comprehensively deepening the reform of the scientific and technological system. China must earnestly implement the decisions and arrangements of the Central Committee of the CPC and the State Council, face the whole world, the base itself on the overall situation, deeply understand and accurately grasp the new requirements of the new normal of economic development and the new trend of scientific and technological innovation at home and abroad, systematically plan a new path for innovative development, take scientific and technological innovation as a guide to explore a new development realm, accelerate the march into the ranks of innovative countries, and speed up the making of China into a global science and technology power.

## (1) Innovation Leads Chinese Manufacturing

As the main body of the national economy, the manufacturing industry is the foundation of a country, the instrument to rejuvenate a nation, and the basis to make a country strong. The real economy is the essential strength of the national economy. China should develop manufacturing, especially advanced manufacturing. Since the 18th National Congress of the CPC, under the guidance of the development concept of innovation, coordination, green, openness, and sharing, China has fully implemented the manufacturing power strategy, intensely promoted the supply side structural reform, focused on innovation drive, quality and efficiency improvement, vigorously developed new technologies, new industries, new formats, and new models, and opened up a unique situation for the development of Chinese manufacturing.

### FAST

The five-hundred-meter aperture spherical radio telescope (FAST), known as The Chinese Eye in the Sky, is the world's largest single-aperture and most sensitive radio telescope. It was completed in Guizhou in September 2016. The FAST project consists of an active reflector system, feed support system, measurement and control system, receiver and terminal, and observation base. As of September 12, 2018, it has discovered 59 high-quality pulsar candidates, of which 44 have been confirmed as newly discovered pulsars.

Panorama of FAST. ( From FAST Engineering Office)

## The World's First Photo Quantum Computer

On May 3, 2017, academician Pan Jianwei of the University of Science and Technology of China announced in Shanghai that a photon quantum computer had been built by the Chinese scientific research team, which demonstrated for the first time greater quantum computing capability than early classical computers. The experimental test shows that the sampling speed of the prototype is at least 24,000 times faster than that of similar experiments in international peers, and its running speed is 10–100 times faster than that of the first vacuum tube computer and the first transistor computer in human history through comparison with the classical algorithm.

The first quantum computer.

## Groundbreaking Achievements in 5G

On April 19, 2016, General Secretary Xi Jinping delivered an important speech at the symposium on network security and informatization, emphasizing that the core technology controlled by others is the biggest hidden danger. To have China's Internet development initiative and ensure Internet security and national security, China must break through the core technology problems and strive to surpass developed countries. The *Thirteenth Five-Year Plan for Science and Technology* mentions that during the Thirteenth Five-Year Plan period, it is necessary to carry out the R&D of key core technologies and international standards of 5G (fifth-generation mobile communication), as well as critical products such as 5G chips, terminals, and system equipment, focus on promoting the construction of 5G technical standards and ecosystems, support the tackling of technical weaknesses such as chips and instruments of 4G enhancement technology, form a complete broadband wireless mobile communication industry chain, maintain the synchronous development with the advanced international level, and promote China's becoming one of the leading countries in the fields of broadband wireless mobile communication technology, standards, industry, services, and applications. With the network integration development as the main thread, China aims to break through the core key technologies such as integrated network networking, ultra-high speed, and ultra-broadband communication and network support, make a number of groundbreaking achievements in chips, complete sets of network equipment, network architecture, etc., deploy next-generation network technologies in advance, to significantly enhance the international competitiveness of the Chinese network industry.

## Vigorously Develop High-End Numerically-Controlled Machine Tools and Robots

The Thirteenth Five-Year Plan for Science and Technology proposes to focus on the development of human-like intelligent technology driven by big data; break through the human-centered theory, methods, and key technologies of human-machine-object integration, and develop relevant equipment, tools, and platforms; achieve significant breakthroughs in human-like intelligence based on big data analysis, realize human-like vision, human-like hearing, human-like language, and human-like thinking, and support the development of the intelligent industry. On July 8, 2017, the State Council issued the *Development Planning for A New Generation of Artificial Intelligence*. It mentions that artificial intelligence is a strategic technology leading the future. China must take a global view, put the development of artificial intelligence at the national strategic level, systematically arrange and actively plan, and firmly grasp the strategic initiative of international competition in the new stage of artificial intelligence development.

China will adhere to the principles that science and technology leads, that arrangements are systematic, that market guides and that source is open, base itself upon the overall national development situation, accurately grasp the global artificial intelligence development trend, find the breakthrough and main direction, comprehensively enhance the basic capacity of scientific and technological innovation, comprehensively expand the depth and breadth of application in key fields, and comprehensively improve the intelligent level of economic and social development and national defense applications.

## The Successful First Flight of COMAC C919

The C919 passenger aircraft (COMAC C919) is the first civil trunk liner developed and assembled by China, per the latest international airworthiness standards, cooperating with enterprises in the United States, France, and other countries. Its development commenced in 2008. For the name COMAC, C is the first letter of China and COMAC. The first nine means everlasting, while 19 suggests its full passenger capacity is 190. COMAC C919 is a landmark project for making China an innovative country, and its body holds completely independent intellectual property rights.

COMAC carries out global collaborative manufacturing, realizes collaborative production of millions of components and collaborative development of airborne systems, and produces COMAC C919 through "Internet + collaborative manufacturing."

## Successful First Flight of Kunlong AG600 On Water

The *Kun* (an enormous legendary fish in Chinese mythology) can soar high in the sky, while the *Long* (a Chinese dragon) dominates the sea. On October 20, 2018, China's self-developed large amphibious aircraft, Kunlong AG600, successfully made its first flight on water at Hubei Jingmen Zhanghe Airport. So far, China's large aircraft have finally realized the goal of conquering both the sky and the sea, and the outline of making China an aviation power has become more apparent.

China-made large amphibious aircraft AG600 made its first flight on water.

On April 26, 2017, the launching ceremony of China's first domestic aircraft carrier PLA Navy Shandong was held in the Dalian Shipyard of China Shipbuilding Industry Group Co., Ltd. The illustration shows that it was slowly launched out of the dock and moored.

## Construction of Ocean Engineering Equipment and High-Tech Ships

On April 26, 2017, the aircraft carrier Shandong (PLA Navy Shandong) was launched. It is 315 meters long and 75 meters wide, with a displacement of nearly 70,000 tons and a maximum speed of 31 knots. It can carry 36 J-15 carrier-based aircraft using a ski jump takeoff mode similar to that of the Chinese aircraft carrier Liaoning. PLA Navy Shandong, developed by China itself, is the first real sense of domestic aircraft carrier.

## Vigorously Develop Advanced Rail Transit Equipment

High-speed trains reflect the level of China's equipment manufacturing industry. They are also hotcakes in constructing the Going Global program and the Belt and Road Initiative and are excellent business cards for China. Since the 18th National Congress of the CPC, China has accelerated the construction of an integrated modern infrastructure network that is moderately advanced, safe, efficient, and interconnected, and a large number of key construction projects related to the national economy and the people's livelihood have been completed and put into operation. A comprehensive transportation network with multiple nodes and full coverage has taken shape. By the end of 2018, China's high-speed rail business mileage has exceeded 29,000 kilometers, more than two-thirds of the world's total. China has built the world's largest high-speed railway network with the fastest operation speed and exclusive intellectual property rights.

## Vigorously Develop Energy-Saving and New Energy Vehicles

China is an integral part of the global automotive market, and there is still much room for growth. Automobile technology is developing in a low-carbon, informationized, and intelligent direction. The Paris Agreement adopted at the UN Conference on Climate Change in Paris on December 12, 2015, is a historic breakthrough of the UN Conference on Climate Change. China has also made a solemn commitment to the world. From the perspective of total carbon emissions, optimizing vehicle emissions and promoting energy conservation and emission reduction are the primary measures of China. In terms of promoting low-carbon automobiles, it is mainly to promote low-carbon energy and low-carbon manufacturing of automobiles by coordinating energy conservation and emission reduction in all aspects of the life cycle of new energy automobiles, including the improvement of vehicle technology and power technology.

## Actively Build New Electric Equipment

### 1) AP1000 power unit

On April 25, 2018, the world's first AP1000 nuclear power unit—Sanmen Nuclear Power Plant Unit 1, Taizhou City, Zhejiang Province, was allowed to charge. On September 21 of the same year, it effectively completed the 168-hour full-power continuous operation assessment, thus deemed ready for commercial operation, also becoming the world's first AP1000 nuclear power unit qualified for commercial operation.

### 2) Hualong One

Hualong One is an advanced million-kilowatt pressurized water reactor nuclear power technology developed by China National Nuclear Corporation (CNNC) and China General Nuclear Power Group based on more than 30 years of nuclear power research, design, manufacturing, construction, and operation experience in China and following the latest safety requirements in China and the world. On April 27, 2019, the first hydraulic test of the global reactor of Hualong One, that is, Unit 5 of CNNC Fuqing Nuclear Power Station, officially commenced, which marks that the unit started a cold functional test 50 days ahead of schedule, totally moving from the installation phase to the commissioning stage.

Hualong One is being capped.

## Improve the Localization Level of Agricultural Machinery Equipment

In 2011, the first Dongfanghong tractor with independent intellectual property rights and the largest power in China, produced by YTO Group Corporation Heilongjiang Modern Agricultural Equipment Base, slowly drove off the assembly line in Qiqihar, marking that China possesses the manufacturing capacity of high-power tractors, thus ending the foreign monopoly of heavy agricultural machinery equipment for China while greatly improving the localization level of agricultural machinery equipment. YTO Group Corporation has accomplished many innovative breakthroughs in China's agricultural machinery technology. While deeply expanding the domestic market, it has also made efforts to expand overseas markets, selling its products to more than 100 countries and regions worldwide. Especially since the Belt and Road Initiative was put forward, YTO Group Corporation is using this new economic channel to introduce more and more Chinese agricultural machinery to the global market.

## China Spallation Neutron Source In Operation

On August 23, 2018, China Spallation Neutron Source (CSNS), a major national science and technology infrastructure, passed the national inspection acceptance and was put into operation. CSNS, one of the first of the twelve major scientific devices built during the Eleventh Five-Year

Plan period in China, is a large research platform for high-tech multidisciplinary applications at the international forefront. The CAS and the Guangdong Provincial People's Government jointly completed the project. After completion, it has become the first in China and the fourth pulsed spallation neutron source in the world.

CSNS will provide an advanced and powerful scientific research platform for China's research in frontier fields such as physics, chemistry, life science, material science, nanoscience, medicine, national defense research, and new nuclear energy development, and fill the gap in the field of pulse neutron application in China.

CSNS Fast Cycle Proton Synchrotron. (Photo credit: Institute of High Energy Physics, CAS)

Bird's eye view of CSNS. (Photo credit: Institute of High Energy Physics, CAS)

The cloned monkeys Zhongzhong and Huahua were taken good care of in the incubator of the non-human primate platform nursery of the Institute of Neuroscience, CAS.

## Progress in Biomedicine

### 1) Birth of cloned monkeys Zhongzhong and Huahua

From November to December 2017, two cloned monkeys were born in China.

Since the birth of the first cloned sheep, Dolly, in 1996, scientists worldwide have cloned cattle, mice, cats, and dogs, among other animals, using somatic cells in 20 years or so. Still, they have not overcome the predicament of cloning non-human primates that are closest to humans. Scientists once generally believed that the current technology could not make it happen. After five years of efforts, Sun Qiang's team from the Institute of Neuroscience, CAS, has successfully broken through this frontier problem in world biology.

### 2) The genetic information of wheat deciphered

On August 16, 2018, the genome map program, with the wheat model variety Chinese Spring as a genetic information reference sequence, was successfully completed, and its findings were published in the journal *Science*. The program sequenced and analyzed the 21 chromosome sequences of wheat, accurately located 107,891 genes, and obtained more than 4 million molecular markers and sequence information affecting gene expression. The International Wheat Genome Sequencing Association organized and implemented the program, founded in 2005. More than 200 scientists from 20 countries and 73 research institutions worldwide participated in mapping the wheat genome.

Wheat, rice, and corn are known as the three major food crops in the world, but their genomes are highly complex and have yet to be deciphered. Therefore, this landmark work once again proves the importance of promoting food security through international cooperation. It sets an example for future large-scale and complex plant genome sequencing and provides a basis for further research on wheat disease resistance and breeding. The completion of the wheat genome map has laid a foundation for breeding wheat varieties with a higher yield, richer nutrition, and stronger climate adaptability. Using this map to improve wheat genetic traits is bound to become one of the important ways to solve the growing demand of the global population for wheat yield.

## (2) Space Power

In the *Thirteenth Five-Year Plan for Science and Technology*, the high-resolution earth observing system, manned space, and lunar exploration projects are listed as major national science and technology projects, and specific goals are clarified. The high-resolution earth observing system is expected to complete the construction of space-based and aerial observation systems, ground systems, and application systems, and basically build land, atmosphere, and ocean earth observation systems and form a complete system. The manned space and lunar exploration projects are expected to launch a new high-thrust launch vehicle, Tiangong-2 space laboratory, space station test core module, manned spacecraft, and cargo spacecraft; master technologies such as cargo transportation and medium and long-term stay of astronauts, to lay the foundation for the comprehensive construction of China's near earth manned space station. Overall, they are expected to make a breakthrough in key technologies such as all moon landings, high data-rate communication, high-precision navigation, and positioning, and lunar resource development, master the automatic return technology of extraterrestrial objects, develop the launch lunar sampling returners, and realize soft landing and sampling return in specific areas.

### Tiangong-2

The Tiangong-2 space laboratory is the second space laboratory independently developed in China after Tiangong-1. It was successfully launched at Jiuquan Satellite Launch Center on September 15, 2008, to verify the space rendezvous and docking technology further and carry out a series of space experiments. On October 17 of the same year, the Shenzhou XI spacecraft took off at the Jiuquan Satellite Launch Center and made rendezvous and docking with Tiangong 2, forming a complex. Subsequently, astronauts were stationed in Tiangong-2, in orbit for 30 days, working and living in accordance with the flight manual, operating instructions, and ground instructions, and conducted relevant scientific experiments as planned.

## New Launch Vehicles

### 1) Long March 7

The Long March 7 launch vehicle is a new type of liquid fuel launch vehicle developed by the China Academy of Launch Vehicle Technology as the overall development unit, as well as a new generation of medium-sized launch vehicles newly developed by the China Manned Space to launch cargo spacecraft. Its predecessor is the Long March 2F launch vehicle. Long March 7 adopts a "two and a half" configuration, with a total length of 53.1 meters, a core diameter of 3.35 meters, and four bundled boosters, each with a diameter of 2.25 meters. Its low Earth orbit carrying capacity is no less than 14 tons, and its 700 km sun-synchronous orbit carrying capacity is 5.5 tons. On June 25, 2016, it took off successfully from Wenchang Satellite Launch Center, which was the first launch mission of Wenchang Satellite Launch Center.

Wenchang Satellite Launch

Long March 5 launch vehicle.

Center is located in Longlou Town, Wenchang City, Hainan Province, China. It is the first coastal launch base in China and one of the few low-latitude launch sites in the world. The launch center, able to launch Long March 5 series rockets and Long March 7 launch vehicles, is mainly responsible for launching geosynchronous orbit satellites, large mass polar orbit satellites, large tonnage space stations, and deep space exploration satellites, among other spacecraft.

### 2) Long March 5

The Long March 5 series launch vehicle is a new generation of launch vehicles with a core diameter of five meters developed by the China Academy of Launch Vehicle Technology. It is a large liquid fuel launch vehicle that is non-toxic, pollution-free, high-performance, low cost, and high thrust. The Long March 5 series launch vehicle is designed to focus on generalization, serialization, and combination. It adopts a "one and a half" or "two and a half" structure, with a carrying capacity of 25 tons in low Earth orbit, and 14 tons in a geosynchronous transfer orbit, in the same class as the European Ariane 5 launch vehicle. On November 3, 2016, Long March 5 was launched at Wenchang Space Launch Site.

### 3) Long March 11

The Long March 11 launch vehicle is a small all-solid fuel launch vehicle developed by the China Academy of Launch Vehicle Technology, China Aerospace Science and Technology Corporation. It has a greater ability to enter space quickly and launch in an emergency. The launch period of Long March 11 is within 72 hours, and its minimum launch time is within 24 hours. Its system consists of a solid launch vehicle and launch support system, with a takeoff thrust of 120 tons. On September 25, 2015, Long March 11 successfully made its first flight at Jiuquan Satellite Launch Center.

On June 5, 2019, China used the Long March 11 sea launch vehicle in the Yellow Sea area to send the technical test satellites Bufeng 1A and 1B and five commercial satellites into their predetermined orbit. The test was a triumph. This is the first time China has performed a launch technology test of a launch vehicle at sea.

## Dark Matter Practice Explorer

On December 17, 2015, China's independently developed dark matter particle explorer (DMPE) was successfully launched. The launch aimed to find a mysterious thing in the universe, dark matter, which is invisible and intangible. On November 30, 2017, *Nature* published the first batch of important achievements obtained by Chinese scientists through the DMPE. Chinese scientists have gradually moved from the learners, successors, and spectators of significant discoveries and theories in the frontier of natural science to the center of the stage.

## First Quantum Communication Satellite

On August 16, 2016, Micius Satellite for Quantum Science Experiments, the world's first quantum experiment satellite led by the University of Science and Technology of China, was successfully launched by a Long March 2D launch vehicle at Jiuquan Satellite Launch Center. China has independently developed it through a series of critical technologies such as satellite platforms, payloads, ground optical transceiver stations, etc. It was expected to carry out scientific experiments such as quantum key distribution, wide area quantum key network, quantum entanglement distribution, quantum teleportation, satellite ground high-speed coherent laser communication, etc., thus enabling China to realize quantum communication between satellites and ground for the first time in the world, and to build a quantum science experiment system that integrates space and earth.

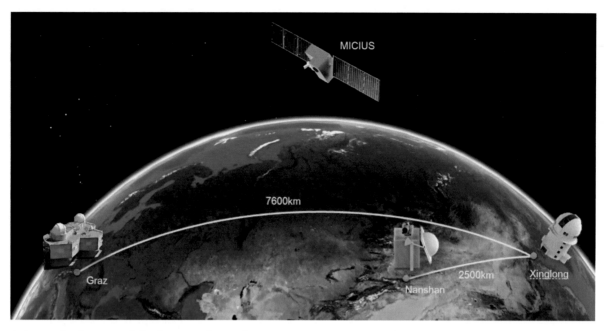

Schematic diagram of the intercontinental quantum-secret communication network.

## Beidou Enters the Era of High-Density Networking

On November 19, 2018, the Long March 3B launch vehicle successfully sent China's 18th and 19th Beidou-3 navigation satellites into space. So far, the deployment of the Beidou-3 basic system constellation has been completed, and Beidou has taken a critical step from regional to global.

On April 20, 2019, the 44th Beidou navigation satellite was successfully launched, the first time that the Beidou-3 system launched an inclined geosynchronous orbit (IGSO) satellite. The Beidou-3 system is a hybrid constellation composed of three different types of orbit satellites, including the MEO (medium Earth orbit) satellite, the GEO (geosynchronous orbit) satellite, and the IGSO satellite. This constellation is also unique to the Beidou system and the first in the world. The IGSO satellite launch can improve the Beidou system's performance in the Asia Pacific region, including its anti-occlusion capability and accuracy, so it performs more accurately in Asia. This launch was the first launch of the Beidou navigation satellite in 2019, which marks the beginning of Beidou's high-density networking.

## Tianzhou-1

Tianzhou-1 cargo spacecraft is the first Chinese cargo spacecraft. It has the functions of rendezvous and docking with the Tiangong-2 space laboratory, on-orbit propellant replenishment, and space science and technology experiments. Tianzhou-1 is a fully sealed cargo spacecraft. It adopts a two-cabin configuration: a cargo cabin and a propulsion cabin. 10.6 meters long, it has a largest diameter of 3.35 meters and a takeoff mass of 12.91 tons. The maximum width of its solar panel once extended is 14.9 meters. Its material transportation capacity is about 6.5 tons, its propellant replenishment capacity is about 2 tons, and it can fly independently for three months.

On April 20, 2017, Tianzhou-1 was sent to space by Long March 7 Yao-2 at Wenchang Satellite Launch Center, and on April 27, it completed the first on-orbit propellant replenishment test with Tiangong-2, marking the triumph of the Tianzhou-1 mission.

## Gaofen 5 and 6 Satellites

At 2:28 on May 9, 2018, China successfully launched Gaofen 5 satellite with a Long March 4C launch vehicle at Taiyuan Satellite Launch Center. Gaofen 5 is the world's first full-spectrum hyperspectral satellite to comprehensively observe the atmosphere and land, and also an important scientific research satellite in China's Gaofen project. It fills the gap where Chinese satellites cannot effectively detect regional air pollution gases, meeting the urgent need for comprehensive environmental monitoring and other aspects. It is an important symbol of China's capability to achieve high spectral resolution earth observation. On June 2, 2018, Gaofen 6 satellite was successfully launched by the Long March 2D launch vehicle at Jiuquan Satellite Launch Center. Gaofen 6 is a low-orbit optical remote sensing satellite and the first high-resolution satellite to achieve precision agricultural observation in China. It is networked with the orbiting Gaofen 1 satellite to significantly improve agriculture, forestry, and grassland monitoring capacity, among other resources.

On May 9, 2018, China successfully launched Gaofen 5 satellite with a Long March 4C launch vehicle at Taiyuan Satellite Launch Center.

## China Seismo-Electromagnetic Satellite

On February 2, 2018, China successfully sent the electromagnetic monitoring test satellite Zhangheng 1 with the Long March 2D launch vehicle from Jiuquan Satellite Launch Center into its scheduled orbit. This marks that China has become one of the few countries worldwide with high-precision geophysical field detection satellites in orbit.

## Chang'e-4 Landed on the Back of the Moon

On December 8, 2018, Chang'e-4 was successfully launched by Long March 3B launch vehicle at Xichang Satellite Launch Center, opening a new journey for China's lunar exploration. On December 30, the probe successfully completed orbit change control in the lunar orbit and entered its scheduled lunar back landing preparation orbit. On January 3, 2019, the Chang'e-4 lander, carrying the Yutu-2 rover, completed the world's first soft landing on the back of the moon.

On May 15, 2019, *Nature* published a major discovery of lunar exploration. The team led by Li Chunlai of the National Astronomical Observatories, CAS, using the spectral exploration data of Chang'e-4, proved that there are deep materials mainly composed of olivine and low calcium pyroxene in the South Pole Aitken basin (SPA) on the back of the moon, providing direct evidence for the composition of the lunar mantle materials, which will support the improvement the model of the formation and evolution of the moon.

## (3) Major Progress in Deep Exploration Technology

On June 2, 2018, Earth Crust I, a Chinese super drilling rig, officially announced the completion of its "debut": set a new record (7,018 meters) for continental scientific drilling in Asian countries, marking that China became the third country in the world, following Russia and Germany, to hold special equipment and related technologies for implementing 10,000-meter continental drilling programs. The development and application of Earth Crust I is an important breakthrough in the independent capacity building of China's deep exploration program. It marks significant progress in China's geoscience field's "into the earth" program for deep exploration of the earth. It provides high-tech means for the comprehensive implementation of subsequent crustal exploration projects and the exploration of deep mysteries of the planet.

Huang Danian (1958–2017) is a geophysicist, former head of the emerging interdisciplinary science department of Jilin University, and professor and doctoral supervisor of the School of Earth Exploration Science and Technology. With the joint efforts of Huang Danian, leader of the Deep Exploration Key Instrument and Equipment R&D and Experiment Project, and his team, China has made remarkable progress in key technologies, such as ultra-high precision mechanical and electronic technology, nano and micro motor technology, high-temperature and low-temperature superconducting principle technology, cold atom interference principle technology, optical fiber technology and inertial technology, and the R&D of rapid mobile platform detection technology and equipment has also overcome the bottleneck for the first time. Huang Danian has led the team to create many of China's firsts, filled a number of technical gaps for its "sky patrol, earth exploration, and sea diving," and made outstanding contributions to the exploration of deep earth resources and the construction of national defense security.

Complete machine system
of the Earth Crust I
10,000-meter drilling rig.

## (4) Actively Build "Internet+"

On April 19, 2016, General Secretary Xi Jinping delivered a speech at the symposium on network security and informatization, requiring to "strengthen the construction of information infrastructure, consolidate the deep integration of information resources, and open up the information artery of economic and social development" and "adapt to people's expectations and needs, accelerate the popularization of information services, reduce application costs, provide people with useful, affordable and good information services, and enable the public to gain more in sharing the fruits of Internet development."

## (5) A Maritime Power

On May 18, 2017, the pilot production of natural gas hydrate (also known as combustible ice) in the Shenhu area of the South China Sea accomplished 187 consecutive hours of stable gas production. This was the first successful trial production of combustible ice in the sea area in China. It is an outstanding Chinese theory, technology, and equipment achievement and has profoundly impacted the energy production and consumption revolution.

Natural gas hydrates are distributed in deep-sea or permafrost regions, and their combustion produces only a small amount of carbon dioxide and water, resulting in much less pollution than coal, oil, and other fossil fuels. Furthermore, they have enormous reserves, so they are internationally recognized as a potential substitute for sources like petroleum.

## (6) Make China Healthy

From August 19 to 20, 2016, the National Conference on Health and Hygiene was held in Beijing. On October 17, the same year, the Central Committee of the CPC and the State Council issued the *Outline of the Healthy China 2030 Plan*, which is the action plan for making China healthy in the next 15 years. It proposes to adhere to the people-centered development idea, firmly establish and implement the development concept of innovation, coordination, green, openness, and sharing, stick to the correct health and health work policy, and the principles of health priority, reform, and innovation, scientific development, fairness, and justice. It takes improving people's health as the core, the reform and innovation of systems and mechanisms as the driving force, and the popularization of a healthy life, the optimization of health services, improvement of health security, the building of a healthy environment, and the development of health industries together as the focus by starting with a wide range of health factors. It integrates health into all policies, ensuring people's health in an all-round and full-cycle manner to improve health standards and health equity significantly.

In July 2019, the State Council issued the *Opinions of the State Council on the Implementation of the Healthy China Action* (after this referred to as the *Opinions*), founded the Healthy China Action Promotion Committee at the national level, and issued the *Healthy China Action (2019–2030)*, which is a new "construction drawing" for further promoting the making of a healthy China. The Healthy China Action adheres to the basic principle of "popularizing knowledge, improving literacy, self-regulation, healthy lifestyle, early intervention, improving services, and public participation, building and sharing together," focuses on preventing and controlling diseases from the source, and on the main problems and critical illnesses that currently affect people's health, and highlights health promotion and mobilization.

The *Opinions* clearly implemented 15 particular actions. Comprehensive measures are to be taken from the aspects of health knowledge popularization, appropriate diet, nationwide fitness programs, smoking control, mental health promotion, etc., to intervene in the health factors comprehensively; it pays special attention to women and children, primary and secondary school students, labor workers, the elderly, among other key groups, to maintain the health of their whole life cycle; it strengthens the prevention and control of major diseases, including cardiovascular and cerebrovascular diseases, cancer, chronic respiratory diseases, diabetes, and infectious and endemic diseases. Relevant special actions also put forward suitable measures for disability prevention and rehabilitation services and health promotion of residents in poor areas.

## (7) Strengthen National Defense

### Xian Y-20

On January 26, 2013, Xian Y-20 heavy transport aircraft independently developed by China conducted a test flight. This type of aircraft is a large, multi-purpose transport aircraft developed by China. It can carry out long-distance air transport tasks of various materials and personnel under complex meteorological conditions. The success of the first flight of the Xian Y-20 is of great significance for promoting China's economic and national defense modernization and responding to emergency situations such as emergency rescue and relief, humanitarian assistance, etc. On July 6, 2016, Xian Y-20 joined the air combat troops.

The Xian Y-20 is a new generation of military large transport aircraft manufactured by China.

## Chengdu J-20

On January 11, 2011, the first demonstration aircraft of the Chengdu J-20 stealth fighter carried out its first flight test. Next, it underwent several more flight tests, whose intensity gradually increased. The Chengdu J-20 adopts a canard aerodynamic configuration with a single seat, double engines, full motion, double vertical tails, DSI bulging inlet, and an upper anti-canard wing with pointed arch strakes. Its head and fuselage are rhombic, its vertical tail wing inclines outward, its landing gear hatch is designed with serrated edges, and its fuselage is painted in bright silver gray. Its side bomb hatch adopts an innovative structure, which can pre-enclose the missile launching rack outside the hatch, and is equipped with new PL-15 and PL-21 air-to-air missiles. On November 1, 2016, the Chengdu J-20 made its first public appearance at the 11th China International Aviation and Aerospace Exhibition. On February 9, 2018, the Chengdu J-20 joined the air combat troops.

The Chengdu J-20 is a stealthy fifth-generation air superiority fighter aircraft with high stealth capabilities, advanced situational awareness, and exceptional maneuverability.

## (8)  Establish a Big Innovation Community

### High Duality Development of the Belt and Road Initiative

On September 7, 2013, President Xi Jinping delivered a speech entitled Promote Friendship Between Our and Work Together to Build a Bright Future at Nazarbayev University in Kazakhstan, proposing jointly building the economic belt along the Silk Road. On October 3, 2013, President Xi Jinping delivered a speech entitled "Join Hands to Build a China-ASEAN Community with a Shared Future" at the Indonesian parliament, proposing jointly building the 21st Century Maritime Silk Road. The economic belt along the Silk Road and the 21st Century Maritime Silk Road are referred to as the Belt and Road initiative.

From May 14 to 15, 2017, the Belt and Road Forum for International Cooperation was held in Beijing. This forum is the highest standard international event under the framework of the Belt and Road Initiative, the highest level and largest multilateral diplomatic event initiated and hosted by China since the founding of the PRC, and an important symbol of China's significant improvement in international status and influence. Heads of state and government from 29 countries attended the event, including more than 1,500 representatives from more than 130 countries and 70 international organizations, covering all major regions on five continents. Through this forum, countries have formed a list of achievements in 5 categories, 76 items, and more than 270 sub-items.

From April 25 to 27, 2019, China hosted the second Belt and Road Forum for International Cooperation in Beijing. Compared with the first forum, this one was larger in scale, richer in content, and had more participating countries and more fruitful results. The theme this time was "Belt and Road Cooperation, Shaping A Bright Shared Future. At the round table summit, the leaders and heads of international organizations had in-depth discussions on topics such as "promote connectivity, tap new growth drivers," "strengthen policy docking, build tighter partnerships," "promote green and sustainable development, and implement the UN 2030 Agenda", improved the concept of cooperation, clarified priorities for cooperation, strengthened cooperation mechanisms, and reached a consensus on high-quality joint construction of the Belt and Road Initiative. This forum has sent a clear signal to the public: there are more partner countries in the Belt and Road Initiative, their quality of cooperation is getting higher and higher, and the development prospect is increasingly brighter. China is willing to work with all parties to implement the consensus of this forum, and in the spirit of Chinese collaborate-style painting, to jointly promote the Belt and Road cooperation to become deep, practical, stable and long-term, high-quality, and create a brighter future.

### Coordinated Development of the Beijing–Tianjin–Hebei Region

Promoting the coordinated development of Beijing, Tianjin, and Hebei is a major decision and deployment made by the Central Committee of the CPC with comrade Xi Jinping at its core in the new era. Beijing, Tianjin, and Hebei, close to the Bohai Sea, lead the Three Norths. They cover an area of more than 200,000 square kilometers and house more than 100 million residents. The coordinated development strategy of Beijing, Tianjin, and Hebei not only aims at treating the "urban disease" in Beijing and overcoming the complex challenge of regional coordinated development but also strives to resolve the serious overload contradiction of resources and environment and explore a high-quality development model in densely populated areas. Overlooking today's Beijing–Tianjin–Hebei region, the new airport in Beijing looks like a golden phoenix soaring high, Baiyang Lake has become a veritable pearl in North China after the ecological improvement, and Tianjin Port's ships are bustling with ships coming and going... Since implementing the coordinated development strategy of Beijing–Tianjin–Hebei, remarkable achievements have been made.

Beijing Daxing International Airport at dawn.

## Construction of Guangdong–Hong Kong–Macao Greater Bay Area

The Guangdong–Hong Kong–Macao Greater Bay Area includes Hong Kong Special Administrative Region, Macao Special Administrative Region, Guangzhou, Shenzhen, Zhuhai, Foshan, Huizhou, Dongguan, Zhongshan, Jiangmen, and Zhaoqing, with a total area of 56,000 square kilometers, and a total population of about 70 million in 2017. It is one of the regions with the highest degree of openness and the strongest economic vitality in China, thus occupying an important strategic position in the overall development of China. The construction of the Guangdong–Hong Kong–Macao Greater Bay Area is not only a new attempt to promote the formation of a new pattern of comprehensive opening up in the new era but also a new practice to promote the development of the cause of "one country, two systems."

*1) Promote economic development and enhance people's well-being*

To fully implement the spirit of the 19th National Congress of the CPC, China fully and accurately executes the "one country, two systems" policy, gives full play to the comprehensive advantages of Guangdong, Hong Kong, and Macao, deepens cooperation between the mainland and Hong Kong and Macao, further enhances the supporting and leading role of the Guangdong–Hong Kong–Macao Greater Bay Area in national economic development and opening up, supports Hong Kong and Macao to integrate into the overall national development, enhances the well-being of Hong Kong and Macao compatriots, and maintains long-term prosperity and stability of Hong Kong and Macao, so that they share with the motherland the historical responsibility for national rejuvenation and the great glory of the motherland's prosperity.

The Guangdong–Hong Kong–Macao Greater Bay Area provides a rare opportunity for innovation and entrepreneurship. Many young and middle-aged entrepreneurs choose to settle down here. More and more SMEs of science, technology, culture, and finance, among other types, have sprouted and expanded, deeply involved in the national boom of rapid development.

The opening of the Hong Kong-Zhuhai-Macao Bridge has dramatically facilitated transportation between the three cities, narrowing the distance between the citizens of the three cities.

*2) Establish a transportation network*

The construction of transportation infrastructure is an important carrier and main content of the construction of the Guangdong–Hong Kong–Macao Greater Bay Area. As the hinterland of the Guangdong–Hong Kong–Macao Greater Bay Area, Guangdong has continued to accelerate the construction of transportation infrastructure in recent years and improve the level of comprehensive transportation services, and gradually formed the external transportation network of the Guangdong–Hong Kong–Macao Greater Bay Area. Currently, the Guangdong–Hong Kong–Macao Greater Bay Area's sea, land, and external air channels have formed a network, with a total volume of passenger transport and freight transport accounting for more than 35% of the national total. It is ready to become an air-land-sea gateway and hub with complete functions, timely and reliable, convenient customs clearance, smooth circulation, and economic efficiency while connecting the Belt and Road Initiative.

*3) Hong Kong–Zhuhai–Macao Bridge opened to traffic*

Hong Kong–Zhuhai–Macao Bridge is another major infrastructure project in China following the Three Gorges Project, the Qinghai–Xizang Railway, the South to North Water Transfer Project, the West to East Gas Transmission Project, and the Beijing–Shanghai High-Speed Railway. Connecting Hong Kong to the east and Zhuhai and Macao to the west, it is a super large sea crossing channel that integrates bridges, islands, and tunnels. In addition to the standard features of super large projects, such as grand scale, tight construction period, serious difficulty, and high risk, the project is also characterized by great social attention, joint construction, and management by the three governments, general contracting mode of design and construction, as well as the restrictions of White Dolphin Reserve, complex navigation environment, construction period and interfaces. Its construction started on December 15, 2009; on July 7, 2017, the main body of the bridge was completed; on February 6, 2018, the main body passed the inspection and acceptance, and a joint test run of Guangdong, Hong Kong, and Macao was conducted on September 28 of the same year. On October 23, 2018, the opening ceremony of the Hong Kong–Zhuhai–Macao Bridge was held in Zhuhai, Guangdong Province. On October 24, it was officially opened to traffic.

# (9) Solidly Improve the Scientific Literacy of the Chinese People

The "Three Conferences on Science and Technology" proposed that scientific and technological innovation and science popularization are the two wings to achieve innovative development. It is necessary to put science popularization in the same important position as scientific and technological innovation, popularize scientific knowledge, promote the scientific spirit, disseminate scientific concepts, advocate scientific methods, and create a good atmosphere of emphasizing science, loving science, learning science, and using science in the whole society, to fully release the innovative wisdom hidden among hundreds of millions of Chinese and make full use of this innovation force.

Since the Thirteenth Five-Year Plan, under the close attention and correct leadership of the Central Committee of the CPC and the State Council, all departments in various regions have made solid progress in promoting the scientific literacy of the Chinese people. Their scientific literacy has rapidly improved. Their public service ability has been significantly stronger. The mechanism for promoting citizens' scientific literacy has been further perfected, and remarkable achievements have been made in science popularization. In 2018, the proportion of Chinese citizens with scientific literacy reached 8.47%, only 1.53%, away from the target of 10% in 2020. The primary conditions for science education and training have been greatly improved, the construction of a modern science and technology museum system has taken great leaps, the role of informatization in science popularization has been increasingly important, and the pattern of wide union and cooperation has been increasingly enhanced.

# Appendix

# Highest Science and Technology Awards

From 1950 to 1966, China issued the *Regulation of the People's Republic of China on Awards for Inventions*, among other important regulations, initially establishing the Chinese science and technology awards system.

Since 1978, China has further reformed and improved its science and technology awards system *the Regulations of the People's Republic of China on Awards for Scientific and Technological Progress* issued in 1984 was the first comprehensive science and technology award regulation in China, while the *Law of the People's Republic of China on Scientific and Technological Progress* issued in 1993 further established the legal status of the science and technology awards system.

During the 20 years from 1979 to 1999, China's science and technology awards system has achieved fruitful results. More than 60,000 Science and Technology Award winners and 12,582 scientific and technological achievements have been rewarded.

In 1999, China reformed its science and technology awards system, canceling the awards set by departments, adjusting the awards settings, and adding the highest science and technology award. The Highest Science and Technology Award is the highest-level science and technology award in China at present, with up to two winners each year, and the head of state presents the award in person. All winners are Chinese citizens who have either made significant breakthroughs at the forefront of contemporary science and technology or created tremendous economic or social benefits in scientific and technological innovation and transformation of scientific and technological achievements.

# Winners of the Year 2000

**Wu Wenjun** (1919.05–2017.05), male, mathematician, academician of the CAS, and researcher of the Academy of Mathematics and Systems Science of the CAS. In the 1950s, he made outstanding achievements in the research of characteristic class and embedding class, such as the Wu Wenjun formula and Wu Wenjun's characteristic class, which had many important applications. In the 1970s, he created the "Wu method" of mechanical theorem-proving in geometry, which has great influence and essential application value, thus transforming the mathematical research methods.

**Yuan Longping** (1930.09–2021), male, hybrid rice expert, academician of the CAE, and researcher at the Hunan Academy of Agricultural Sciences. In 1964, he began to study hybrid rice. In 1973, he realized a three-line combination. In 1974, he bred the first strong hybrid rice combination Nanyou No .2. In 1975, he successfully developed hybrid rice seed production technique, which laid a foundation for the wide promotion of hybrid rice. In 1985, he proposed the strategic vision of hybrid rice breeding, illuminating the direction for developing hybrid rice. In 1987, he served as the expert on the two-line hybrid rice program of Project 863. In 1995, he succeeded in developing two-line hybrid rice. In 1997, he proposed the technical route of super hybrid rice breeding. In 2000, he achieved the first phase goal of China's super rice breeding set up by the Ministry of Agriculture, and in 2004, he completed the second phase goal of super rice a year ahead of schedule.

# Winners of the Year 2001

**Huang Kun** (1919.09–2005.07), male, physicist, academician of the CAS, and researcher at the Institute of Semiconductors, CAS. In 1950, Huang first proposed the quantum theory of multi-phonon radiative and non-radiative transitions. In 1951, he first proposed the coupled oscillation mode of phonons and electromagnetic waves in crystals, which was verified by the international Raman scattering experiment in 1963. It was named as a kind of elementary excitation—polariton and the equation of motion he proposed was internationally known as the "Huang equation." Over a decade or so, he made new achievements, in cooperation with young colleagues, in the theory of multi-phonon transitions and quantum well superlattices. With him as the academic leader, the Institute of Semiconductors has established The State Key Laboratory of Superlattices and Microstructures in China, which has initiated and developed China's research work in this new field of materials science and solid-state physics.

**Wang Xuan** (1937.02–2006.02), male, computer expert, academician of the CAS, academician of the CAE, and professor at Peking University. Before 1975, he was engaged in the research of computer logic design, architecture, and high-level language compilation systems. In 1975, he led the development of Huaguang and Founder computer laser Chinese editing-phototypesetting system, which is used for editing books, newspapers, and other official publications. The Huaguang and Founder systems developed under his leadership have gradually become popular in Chinese newspapers, publishing houses, and printing plants, laying a foundation for the computerization of the whole process of news publishing.

# Winners of the Year 2002

**Jin Yilian** (1929.09–), male, computer expert, academician of the CAE, and researcher at the State Research Center for Parallel Computer Engineering Technology. As one of those in charge of the operation control department, he participated in developing the first general-purpose large electronic computer in China. From then on, he has long been committed to the research and practice of electronic computer architecture, high-speed signal transmission technology, and computer assembly technology, among other aspects. He led the development of several leading large computer systems in China at that time, one after another. During this period, he proposed specific design plans, made many critical decisions, solved many complex theoretical and technical problems, and made outstanding contributions to the development of China's computer industry, especially parallel computer technology.

# Winners of the Year 2003

**Liu Dongsheng** (1917.11–2008.03), male, an expert in global environmental science, an academician of the CAS, and a researcher at the Institute of Geology and Geophysics, CAS. He was engaged in geoscience research for nearly 60 years. He made many original research achievements in China's paleovertebrate zoology, quaternary geology, environmental science, and environmental geology, Qinghai-Xizang Plateau, and polar survey, among other scientific research fields, especially in loess research, enabling China to rank among the world's top in the field of paleoglobal change studies

**Wang Yongzhi** (1932.11–), male, space technology expert, academician of the CAE, and researcher of the General Reserve Department of the Chinese People's Liberation Army. He has been engaged in space technology since his return to China in 1961. Since 1992, he has been the chief designer of

China's manned space programs. He is one of the pioneers of China's manned space programs and its academic and technical leader. Over the past 40 years, he has made outstanding contributions to developing China's strategic rockets, ground-to-ground tactical rockets, and launch vehicles, especially to manned space projects.

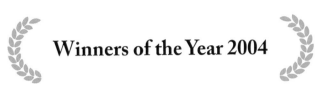

# Winners of the Year 2004

None.

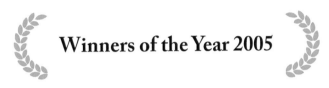

# Winners of the Year 2005

**Ye Duzheng** (1916.02–2013.10), male, meteorologist, academician of the CAS, and researcher of the Institute of Atmospheric Physics, CAS. He used to study atmospheric circulation and long wave dynamics in the early stage. Following C.G. Rossby, he proposed the theory of energy dispersion of long waves, which is an essential contribution to dynamic meteorology. In the 1950s, he and climatologist Hermann Flohn both suggested that the Qinghai-Xizang Plateau is a heat source in summer, which opened up the research on the thermal effect of large terrain. In 1958, Tao Shiyan and others proposed the abrupt seasonal change of the atmospheric circulation in the Northern Hemisphere, leading to a series of studies on this topic. In the 1960s, Ye made important contributions to the adaptive theory of atmospheric wind and pressure fields. From the late 1970s, he was engaged in researching the earth's atmosphere relationship and global changes, enabling China's research in this field to occupy a seat in the world. He was the chief scientist of the state's major basic research project, the Eighth Five-Year Plan, Prediction of the Trend of Changes in China's Living Environment in the Future (20–50 Years).

**Wu Mengchao** (1922.08–2021.05), male, expert in hepatobiliary surgery, academician of the CAS, professor at the Naval Medical University, and one of the main founders of hepatobiliary surgery in China. In the 1950s, he was the first to propose the new "five-lobe and four-segment" theory for Chinese liver anatomy; in the 1960s, he pioneered liver resection in normothermic hepatic intermittent ischemia and took the lead in breaking through the forbidden zone of the middle lobe of the liver; in the 1970s, he established a complete early diagnosis and treatment system for cavernous hepatic hemangioma and small liver cancer and was one of the earlier doctors to apply hepatic artery ligation and hepatic artery embolization to treat middle and late stage liver cancer; in the 1980s, he invented blood free hepatectomy at room temperature, resection of recurrent liver cancer and secondary surgery for liver cancer; in the 1990s, he made significant progress in gene immunotherapy, liver cancer vaccine, liver transplantation, among other aspects of advanced liver cancer, and was the first to perform laparoscopic hepatectomy and hepatic artery ligation.

 **Winners of the Year 2006**

**Li Zhensheng** (1931.02–), male, expert in wheat genetics and breeding, academician of the CAS, and researcher of the Institute of Genetics and Developmental Biology, CAS. He has bred new hybrid types, such as octoploid trititrigia, alien addition line, alien substitution line, and alien line, were bred; he has transferred the excellent genes of drought resistance, resistance to dry and hot wind, and resistance to various wheat diseases of Elytrigia repens to wheat. He has bred new varieties of trititrigia No. 4, 5, and 6. By 1988, trititrigia No. 6 had been promoted in an area of 54 million mu, increasing wheat production by 1.6 billion kilograms; he had created a new wheat chromosome engineering breeding system. Using the blue endosperm gene of Elytrigia repens as a genetic marker character, he was the first to develop a blue grain monosomic wheat system for the first time, which solved the two problems of "univalent chromosome drift" and "too much work for chromosome number identification" that had long existed in the utilization of monosomic wheat; he has bred self-flowering and fruiting nullisomic wheat and has created a new method for rapid breeding of wheat alien substitution lines by using his nullisomic wheat—the nullisomic backcross method, which has laid a foundation for wheat chromosome engineering breeding.

 **Winners of the Year 2007**

**Min Enze** (1924.02–2016.03), male, petrochemical engineering expert, academician of the CAS, academician of the CAE, professor-level senior engineer of Sinopec Research Institute of Petroleum Processing. In the 1960s, he succeeded in developing a phosphoric acid diatomite composite catalyst, platinum reforming catalyst, small ball silica-alumina cracking catalyst, and tiny ball silica-alumina cracking catalyst, and all were put into production. In the 1970s and 1980s, he led the research, development, production, and application of molybdenum nickel phosphorus hydrogenation catalyst, carbon monoxide combustion promoter, semi-synthetic zeolite cracking catalyst, etc. Since 1980, he has led the basic oriented research of new catalytic materials and new chemical reaction engineering, including amorphous alloys, supported heteropoly acids, nanomolecular sieves, magnetically stabilized fluidized beds, suspended catalytic distillation, etc., and has successfully developed new processes such as caprolactam magnetically stabilized fluidized bed hydrogenation and suspended catalytic distillation alkylation. In the 1990s, he led the "Environmentally Friendly Petrochemical Catalytic Chemistry and Reaction Engineering," a major basic research project of the National Natural Science Foundation during the Ninth Five-Year Plan period, entering the field of green chemistry and directing the development of complete green manufacturing technology for chemical fiber monomer caprolactam, which has been industrialized and achieved significant economic and social benefits.

**Wu Zhengyi** (1916.06–2013.06), male, botanist, academician of the CAS, and researcher at Kunming Institute of Botany, CAS. He has expounded and proved the three historical sources and 15 geographical elements of the flora of China. He proposed that the southern and southwestern parts of China between 20° and 40° N be the key areas for the occurrence and development of the flora of the Palaearctic South Continent, the Paleo Arctic North Continent, and the Paleo Mediterranean Sea; *Chinese Vegetation*, which he acted as the chief compiler of, is an essential scientific material for botany-related disciplines and agricultural, forestry and animal husbandry production; he organized and led the investigation of plant resources in China, especially in Yunnan, and pointed out that the formation of valuable substances in plants is related to the distribution area and formation history of plant provenances; he acted as the chief compiler of several national and regional flora records.

# Winners of the Year 2008

**Wang Zhongcheng** (1925.12–2012.09), male, neurosurgery expert, academician of CAE, and professor at Beijing Neurosurgical Institute and Beijing Tiantan Hospital affiliated with Capital Medical University. In the 1950s, he developed new techniques of cerebral angiography in China, which improved the diagnostic rate of intracranial lesions. In 1965, he published the first monograph on neurosurgery—*China Cerebral Angiography*, thus promoting the development of neurosurgery in China. In the 1970s, he started the surgical treatment of cerebrovascular diseases in China, making achievements in the treatment of ischemic cerebrovascular diseases, surgical resection of giant aneurysms and multiple aneurysms, and comprehensive treatment of cerebrovascular malformations. From the 1980s, he devoted himself to studying the treatment of brain stem tumors, a forbidden surgical zone. Then he studied tumors in the spinal cord and successfully performed surgical removal. The two treatments have reached the advanced international level regarding the number of cases, surgical techniques, and results.

**Xu Guangxian** (1920.11–2015.04), male, chemist, academician of the CAS, and professor at Peking University. He was long engaged in the teaching and research of physical chemistry and inorganic chemistry for a long time, involving quantum chemistry, chemical bond theory, coordination chemistry, extraction chemistry, nuclear fuel chemistry, and rare earth science. He proposed a more universal (nxcπ) scheme based on extensive literature, a new concept of atomic covalence, and its quantum chemical definition. According to the molecular structure formula, the stability of metal-organic compounds and cluster compounds can be inferred. He established the quantum chemical calculation method for studying rare earth elements and the chemical bond theory of inorganic conjugated molecules. He synthesized a series of tetra-nuclear rare earth dioxides with special structures and properties. He made many research achievements in the theory of cascade extraction, the law of synergistic extraction, the research methods of extraction mechanism, the technology of rare earth extraction separation, etc.

 **Winners of the Year 2009**

**Gu Chaohao** (1926.05–2012.06), male, mathematician, academician of the CAS, and professor at the Institute of Mathematics of Fudan University. He was engaged in the research and teaching of partial differential equations, differential geometry, mathematical physics, etc. He made important systematic research findings in general space differential geometry, homogeneous Riemannian spaces, infinite-dimensional transformation quasi groups, hyperbolic and mixed type partial differential equations, gauge field theory, harmonic mapping, soliton theory, etc. In particular, he was the first to propose the system theory of high-dimensional and high-order mixed equations and has made significant breakthroughs in the mathematical problems of supersonic flow, the mathematical structure of gauge field, wave mapping, and the study of solitons in high-dimensional space-time.

**Sun Jiadong** (1929.04–), male, space technology expert, academician of the CAS, and senior technical consultant of China Aerospace Science and Technology Corporation. He has been long engaged in the R&D of launch vehicles and man-made satellites. He undertook the overall design of China's first self-designed middle-short-range missile and middle-long-range missile as the overall chief designer; he participated in and led the development and launch of the first artificial earth satellite and recoverable remote sensing satellite; he was the chief technical director and chief designer of various types of satellites; he was responsible for the technical decision-making, command, and coordination of the general system of the moon circling project as the chief designer.

# Winners of the Year 2010

**Shi Changxu** (1920.11–2014.11), male, materials scientist, academician of the CAS, academician of the CAE, specially invited consultant of the NSFC, and researcher of the Institute of Metals of the CAS. As one of the pioneers of Chinese high superalloy, he developed China's first iron-base high superalloy and led the development of China's first generation hollow air-cooled cast nickel-base superalloy turbine blades, which can be used as heat-resistant and low-temperature materials and magnet free manganese aluminum austenitic steel, which is groundbreaking. He participated in or led the formulation of China's science and technology development plan for metallurgical materials, material science, and new materials many times and led the establishment and evaluation of state key laboratories, national engineering research centers, and national major scientific projects.

**Wang Zhenyi** (1924.11–), male, hematologist, academician of CAE, tenured professor at Ruijin Hospital affiliated to the Shanghai Jiaotong University School of Medicine. Since 1954, he has been engaged in the research of thrombosis and hemostasis, and was the first to establish diagnosis methods for hemophilia A and B and mild hemophilia in China. In 1980, he began to study differentiation therapy for cancer. In 1986, he was the first internationally initiated application of all-trans-retinoic acid to induce differentiation in the treatment of acute promyelocytic leukemia, and he achieved a high remission rate, providing a successful example for the new theory that malignant tumors can be treated effectively through induction differentiation therapy without damaging normal cells.

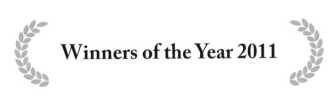

# Winners of the Year 2011

**Xie Jialin** (1920.08–2016.02), male, physicist, academician of the CAS, and researcher of the Institute of High Energy Physics, CAS. He was mainly engaged in accelerator R&D. During his stay in the United States, he led the development of the world's highest energy medical electron linear accelerator. In 1964, he led the construction of China's earliest electron LINAC capable of developing high energy. In the 1980s, he led the design, development, and construction of the Beijing Electron Positron Collider project. In the early 1990s, he directed and built the Beijing Free Electron Laser device.

**Wu Liangyong** (1922.05–), male, architect, academician of the CAS, academician of the CAE, director of the Institute of Architecture and Urban Studies, Tsinghua University, and Director of the Human Settlements Research Center. He has made outstanding contributions in the field of architectural education and has won many awards at home and abroad. In 1996, he was awarded the UIA (Union International des Architectes) Prize for Architectural Criticism or Architectural Education. In addition, he headed and participated in many significant projects, such as the design of the new Beijing Library, the planning and design of the expansion of Tiananmen Square, the planning of the central area of Guilin, Guangxi, the planning and design of the Central Academy of Fine Arts campus, the planning and design of the Confucius Institute, etc. Among them, the pilot project of reconstruction of dilapidated houses in Ju'er Hutong, Beijing, which he led, won the gold medal and World Habitat Award of the ARCASIA Awards for Architecture in 1992.

 # Winners of the Year 2012

**Zheng Zhemin** (1924.10–2021.08), male, mechanics scientist, academician of the CAS, academician of the CAE, and a researcher at the Institute of Mechanics, CAS. In his early years, he studied elastic mechanics, hydro-elastic mechanics, vibration, and earthquake engineering mechanics. Since 1960, he has been engaged in research on explosive processing, underground nuclear explosion, armor piercing, dynamic mechanical properties of materials, volatile treatment of soft underwater foundations, etc. He conducted theoretical research and experiments on explosive forming model law, forming mechanism, die strength, dynamic mechanical properties of explosively formed materials, explosive load, etc. At the same time, he solved the problems of developing parameters and technology, opened up a new direction of "process mechanics" that combines mechanics and technology, and contributed to the theory and application of explosive mechanics.

**Wang Xiaomo** (1938.11–), male, radar technology expert, academician of the CAE, and researcher at the China Academy of Electronics and Information Technology. He has designed and developed various types of radars and systems. The JY-8 radar designed under his leadership has become China's first complete and fully automated three-dimensional radar. It is the cornerstone of a new generation of radar and has been developed into a new radar equipment series. He designed the first JY-9 radar with both high and low altitude in China, which has strong anti-jamming performance and ranks among the best in foreign military exercises and comprehensive scores, becoming one of the best low altitude radars in the world. In 1996, he served as the Chinese chief designer of the Sino-Israeli cooperative airborne warning and control system (AWACS), the general consultant of AWACS after the establishment of domestic AWACS projects in 2000, and the chief designer of self-developed export AWACS. He is the founder of the AWACS project. He is studying space-based information systems and integrated space-earth information networks.

# Winners of the Year 2013

**Zhang Cunhao** (1928.02–), male, physical chemist, academician of the CAS, and researcher at the Dalian Institute of Chemical Physics, CAS. In the 1950s, he studied the water gas catalytic synthesis of liquid fuels. He contributed to developing molten iron catalysts and the solution of heat transfer and anti-mixing problems in fluidized beds. In the 1960s, he devoted himself to studying the combustion of solid rocket propellants and motors, participated in putting forward the burning rate theory and erosion combustion theory, and conducted research on the high-speed reaction kinetics of shock tubes. In the 1970s, he led chemical laser research and developed combustion-driven continuous-wave hydrogen fluoride and deuterium fluoride chemical lasers. In the 1980s and 1990s, he studied the new system of short wavelength chemical lasers and chemical oxygen iodine lasers, spectroscopy, and energy transfer of excited molecules, and then took part in the design of the double resonance ionization method. He studied the accurate energy transfer laws of the rotational energy level structure of ultra-short-lived molecules and the resolution of sub-rotational energy levels of molecular electronic states.

**Cheng Kaijia** (1918.08–2018.11), male, physicist, academician of the CAS, and researcher of the General Reserve of the Chinese People's Liberation Army. As one of the pioneers of Chinese nuclear weapons research, he made outstanding contributions to developing and testing nuclear weapons. He initiated, planned, and led research in new fields of radiation hardening. He was one of the pioneers in the new directional high-power microwave research field in China. He published the first monograph on solid-state physics in China, put forward the general theory of thermodynamic internal friction, derived the Dirac equation, and proposed and developed the superconducting double belt theory and the condensed TFDC electronic theory.

# Winners of the Year 2014

**Yu Min** (1926.08–2019.01), male, nuclear physicist, academician of the CAS, and researcher at the Chinese Academy of Engineering Physics. In the breakthrough of the hydrogen bomb principle in China, he solved a series of fundamental problems, and assumed basic integrity from principle to configuration, thus playing a key role. After that, he long led the theoretical research and design of nuclear weapons and solved extensive theoretical problems. He significantly contributed to further developing China's nuclear weapons to the advanced international level. From the 1970s, he played an essential role in advocating and promoting the research of several high-tech projects.

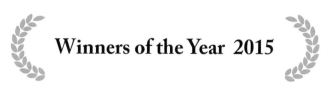

# Winners of the Year 2015

None.

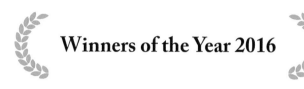

# Winners of the Year 2016

**Zhao Zhongxian** (1941.01–), male, physicist, academician of the CAS, and researcher of the Institute of Physics, CAS. He has long been engaged in low-temperature and superconducting research. From 1967 to 1972, he participated in several national defense scientific research missions. In 1976, he began to explore high-temperature superconductivity. His published papers include the magnetic flux pinning and critical current problems of type II superconductors; and the superconductivity of amorphous alloys. In 1983, he began to study the oxide superconductor BPB system and heavy fermion superconductivity. At the end of 1986, he discovered the influence of impurities in the study of the Ba-La-Cu-O system, and in early 1987, he and his partners independently discovered the yttrium-barium-copper oxide superconductor with a critical temperature of 92.8K.

**Tu Youyou** (1930.12–), female, pharmacist and researcher at the Chinese Academy of Chinese Medical Sciences. After decades of research on traditional Chinese medicine, she and her research team have innovatively developed new antimalarial drugs: artemisinin and dihydroartemisinin. For the discovery of artemisinin, a drug used to treat malaria, she has saved millions of lives worldwide, especially in developing countries. Tu won the Nobel Prize in Physiology or Medicine in 2015.

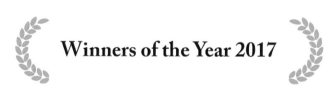

# Winners of the Year 2017

**Wang Zeshan** (1935.09–), male, dynamite scientist, academician of the CAE, professor at Nanjing University of Science and Technology. He has been engaged in teaching and scientific research on energetic materials. He has studied the theory of propellant and its charge; his invention of low-temperature sensing technology improves the launching efficiency and enables the launching power to outperform similar foreign equipment; he has studied and mastered the relevant theory and comprehensive treatment technology of waste explosives and propellants reuse, thus realizing resource reuse, improving safety, reducing public hazards, and generating obvious social and economic benefits; he has invented a kind of high-density propellant charging technology and applied it widely.

**Hou Yunde** (1929.07–), male, molecular virologist, academician of the CAE, researcher of the Institute of Viral Disease Control and Prevention, Chinese Center for Disease Control and Prevention. He has made outstanding achievements in the R&D of molecular virology, genetic engineering interferon, and other genetic drugs, as well as the control of emerging infectious diseases. He has significantly contributed to the industrialization of medical molecular virology, genetic engineering disciplines, biotechnology, and infectious disease control in China. During the new H1N1 influenza pandemic in 2009, as the director of the expert committee of the joint prevention and control mechanism, together with famous scientists in China, with national support, he made collaborative innovation, succeeding in the first human intervention in the influenza pandemic in human history, thus receiving international recognition.

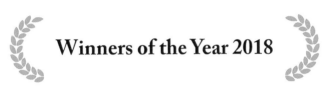

# Winners of the Year 2018

**Liu Yongtan** (1936.12–), male, expert in radar and signal processing technology, academician of the CAS, academician of the CAE, professor at the Harbin Institute of Technology. He developed a new system of sea detection radar, broke through 11 key technologies, solved the problems of signal processing and target detection under the background of strong sea clutter, radio interference, and atmospheric noise, and built China's first new radar station system. In the research of inverse synthetic aperture radar, he has developed the motion compensation theory and proposed a new compensation theory for large bandwidth signals and systems. In developing the digital wideband FM/CW radar signal processor, he put forward a unique analog/digital mixed signal processing mode of microprogrammed analog sliding window segmented FFT spectrum analysis and digital multi-threshold automatic detection, which solves the problem of high-resolution spectrum analysis of large dynamic wideband signals.

**Qian Qihu** (1937.10–), male, a protection engineer, academician of the CAE, and professor at the Army Engineering University of PLA. He has long been engaged in the teaching and scientific research of protection engineering and underground engineering, solved difficult calculation and design problems such as orifice protection, and applied operational research and system engineering methods to the protection engineering field. He led the blasting of Zhuhai Artillery Mountain with the largest charge in the world; he has conducted deep research on deep rock mechanics, protection of deep underground engineering, and development and utilization of underground space; he has led and participated in the design scheme review and bid evaluation of several major domestic metro projects, urban underwater tunnels, and submarine tunnels, and assisted in the construction of Nanjing Yangtze River Tunnel, Shanghai Yangtze River Tunnel and Wuhan Yangtze River Tunnel as the director and member of the expert committee. He has led and completed a number of national consulting topics such as Strategies and Countermeasures for the Development of China's

Urban Underground Space in the 21st Century and Protective Measures and Countermeasures for China's Important Economic Targets; as the chief scientist, he led and completed the research of the NSFC's major project Basic Research and Application of Deep Rock Mechanics.

# References

Cao, Lianqing. "The Birth of the First Ten-Thousand-Ton Hydraulic Compressor in China." *Struggle*, no. 3 (2017): 63.

Dalu, "Introduction to Dongfanghong-150-1 Small Four-Wheel Tractor." *Tractor*, no. 2 (1985): 62–63.

Deng, Nan. *The Development of Science and Technology in People's Republic of China: 1949–2009*. Beijing: China Science and Technology Press, 2009.

Ding, Zhaojun. "Zhang Wenyu, the Founder of Cosmic Ray Physics and High-Energy Physics in China." *Physics Bulletin*, no. 3 (2015): 115–118.

Editorial Committee of the Development of Complete Sets of Equipment for Ten-Million-Ton Large Open Mines. *History of China's Major Technical Equipment: The Development of Complete Sets of Equipment for Ten-Million-Ton Large Open Mines*. Beijing: China Electric Power Press, 2012.

Feng, Lisheng. "Development and Inheritance: Research on the Mechanical History of Liu Xianzhou and Tsinghua University." *Studies in the History of Natural Sciences*, no. 2 (2017): 29–41.

Fu, Yifei. "The Successful First Flight of the Long March 6 Carrier Rocket." *Science and Technology Daily*, September 21, 2015.

Gu, Chaohao. "Comments on the Second Congress of the World Federation of Scientific Workers." *Chinese Science Bulletin*, no. 8 (1951): 61–65.

Gu, Fangzhou. "How to Prevent Polio." *Journal of Nursing*, no. 5 (1965): 55–56.

Guo, Liang, and Zhang Li. "5G Opens a New Era of Interconnection of Everything." *China Science Daily*, June 10, 2019.

*Henan Daily*. "The Good News of the Yellow River Construction Kept Coming—The Red Flag Canal of the Yellow River Completed and Operating." *Yellow River Construction*, no. 10 (1958): 69–70.

Hua, Gong. "Record of Great Projects—A Brief Review of Reports before and after the Yangtze River Bridge Was Completed and Opened to Traffic." *News Business*, no. 11 (1957): 86–88.

Jiang, Zemin. "Speech at the Conference Commending the Scientists and Technicians Who Have Made Outstanding Contribution to the Development of 'Two Bombs and One Satellite.'" *Science News*, no. 28 (1999): 4.

Li, Anping. "A Milestone in the History of the Development of Science and Technology in People's Republic of China—A Twelve-Year Long-Term Plan for the Development of Science and Technology." *Science News*, no. 28 (1999): 32.

Li, Xiang, Wang Xiaoyi, and Cui Junkai. "Zhang Wenyu and His Scientific Contribution—Commemorating the 100th Anniversary of Zhang Wenyu's Birth." *Physics Teacher*, no. 8 (2010): 39–41.

Li, Yunhai. "A Triumph on the Yangtze River—A Review of the Live Broadcast of the Completion and Opening of Wuhan Yangtze River Bridge." *The Press*, no. 2 (1958): 52–53.

Li, Zao. "Stick to the Economic Construction as the Center—A Summary of Studying Comrade Deng Xiaoping's Speech during His Southern Tour." *Research on Mao Zedong and Deng Xiaoping Theory*, no. 3 (1992): 68–72.

Liu, Guangyi, and Chen Zhuo. "Latest Progress of China's 5G Development." *Modern TV Technology*, no. 11 (2018): 55–59.

Lv, Baocheng. "Progress in Smallpox Control in Recent Years (Summary)." *Shanxi Medical Journal*, no. 1 (1963): 89–92.

Ma, Haishan, Qu Xiaoming, and Ye Xin. "Observation on the Therapeutic Effect of Atomized Inhalation of Interferon α1b Injection on Respiratory Syncytial Virus Pneumonia in Children." *Modern Medicine Journal of China*, no. 7 (2019): 81–82.

Ni, Weibo. "HXMT: Space Survey Monitoring Refreshes Human Cognitive Limits." *Science News*, no. 18 (2015): 38–40.

Qian, Yingqian, and Wang Yahui. *Chinese Academic Canon in the 20th Century: Biology*. Fuzhou: Fujian Education Press, 2004.

Shao, Wen. "Introduction to Chen's Theorem—About Chen Jingrun's New Contribution to the Number Theory." *Breaking and Founding (Natural Science Edition)*, no. 3 (1977): 65–68, 72.

Tang, Bo. "Li Siguang's Geological Life." *Map*, no. 6 (2011): 108–115.

Tao, Jun. "The Dream of a Powerful Country in the Deep Ocean—China's First 4,500-Meter-Level ROV Haima (Sea Horse)." *Scientific and Cultural Popularization of Land and Resources*, no. 1 (2016): 15–21.

The Editorial Department. "Accelerate 5G Development in China." *Office Automation*, no. 2 (2018): 32.

The Editorial Department. "Celebrating the Establishment of the Academic Divisions of the Chinese Academy of Sciences." *Chinese Science Bulletin*, no. 6 (1955): 4–6.

The Editorial Department. "The Celebration of the Establishment of the Beijing Center of the World Federation of Scientific Workers and the Preparatory Meeting for the 1964 Beijing Science Symposium." *Chinese Science Bulletin*, no. 11 (1963): 66–67.

The Editorial Department. "China in the Era of Atomic Energy, the First Atomic Reactor Built with the Help of the Soviet Union, and the Cyclotron and the High Pressure Electrostatic Accelerator Began to Work at the Same Time." *China State Farm*, no. 8 (1958): 21.

The Editorial Department. "China in the Era of Satellite Microwave Remote Sensing Applications." *Defense Science & Technology Industry*, no. 9 (2016): 13–14.

The Editorial Department. "China Successfully Launched the High Score Four Satellite." *Aerospace Return and Remote Sensing*, no. 6 (2015): 25.

The Editorial Department. "China's First X-ray Astronomical Satellite Huiyan (Insight) in Use." *Metrology & Measurement Technology*, no. 1 (2018): 57.

The Editorial Department. "The Completion of North China Pharmaceutical Factory—A Great Joy for the Chinese People." *Chinese Pharmaceutical Journal*, no. 4 (1958): 39.

The Editorial Department. "Dayang No.1 Scientific Research Ship Returns to Qingdao." *Marine Geology and Quaternary Geology*, no. 4 (1999): 114.

The Editorial Department. "December 1953: Angang Steel Held the Grand Commencement Ceremony of the 'Three Major Projects.'" *General Review of the Communist Party of China*, no. 12 (2014): 2, 59.

The Editorial Department. "Hard X-ray Modulation Telescope (HXMT) Satellite and Its Scientific Achievements." *Spacecraft Engineering*, no. 5 (2018): 2–12.

The Editorial Department. "July 1956: Changchun FAW Jiefang Automobile Assembled." *General Review of the Communist Party of China*, no. 7 (2014): 2, 59.

The Editorial Department. "The People's Victory Canal Discharges Water." *New Yellow River*, no. 4 (1952): 44.

The Editorial Department. "The Preparation of the Chinese Intellectuals for the Great Advance in Science Under Way." *Philosophical Research*, no. 1 (1956): 148.

The Editorial Department. "Sichuan–Xizang Highway and Qinghai–Xizang Highway." *China's Ethnic Groups*, no. 1 (1965): 57.

The Editorial Department. "The Successful Explosion of China's First Hydrogen Bomb on June 17, 1967." *Trade Union Information*, no. 17 (2015): 39.

Wang, Fan. "Zhou Enlai, Li Fuchun and the Founding of the People's Republic of China's Aerospace Industry—An Interview with Duan Zijun, the Former Deputy Secretary of the Party Leadership Group of the Ministry of Aerospace Industry." *General Review of the Communist Party of China*, no. 4 (2001): 12–19.

Wang, Lihua, and Xu Wei. "Tracing the Breeding Process of Lu Cotton No.1 from the Dusty Technical Records." *China Cotton*, no. 8 (2014): 46–47.

Wang, Songqiao. "The Birth of Chinese First-Class Large Electron Microscope." *Physics Bulletin*, no. 11 (1965): 50–51.

Wang, Zhende. *Modern Science Encyclopedia*. Guilin: Guangxi Normal University Press, 2006.

Wu, Jun, Deng Qiyun, Yuan Dingyang, and Qi Shaowu. "Research Progress in Super Hybrid Rice." *Chinese Science Bulletin*, no. 35 (2016): 65–74.

Xi, Jinping. "Speech on the 40th Anniversary Celebration of the Reform and Opening Up." *Qianjin*, no. 1 (2019): 5–12.

Xu, Bingjin, and Ouyang Min. *The History of Chinese Automobile*. Beijing: China Machine Press, 2017.

Xu, Guangxian. "Structural Rules of Atomic Clusters and Related Molecules I. (nxc π) Scheme." *Huaxue Tongbao*, no. 8 (1982): 44–45.

Xue, Shuixing, Wu Shuhua, Han Feng, Wei Jia, and Hou Yunde. "Improving the Expression of Type α1b Genetically Engineered Interferon in Escherichia Coli." *Chinese Journal of Biotechnology*, no. S1 (1996): 67–71.

Yan, Jinding. "The Current Situation and Strategic Thinking on the Development of Nanoscience and Technology in China." *Chinese Science Bulletin*, no. 1 (2015): 40–47.

Yang, Jingjing. "5G Development Overview and Risks Facing China." *China Information Security*, no. 5 (2018): 16–17.

Yellow River Conservancy Commission of the Ministry of Water Resources. *Sixty Years of People's Governance of the Yellow River*. Zhengzhou: Yellow River Conservancy Press, 2006.

Zeng, Chaoqun, and Zhang Jingcheng. "Design of Democracy No. 10 and Democracy No. 11 for Passenger and Cargo Transportation in Small Ports." *China Shipbuilding*, no. 3 (1957): 3–27.

Zhang, Bochun. "Two Disciplinary Development Plans for the History of Science and Technology in the 1950s—The Draft of the Twelve-Year Long-Term Plan for the Research of the History of Natural Science and Technology in China (1956) and the Outline (Draft) of the Research and Development of the History of Natural Science from 1958 to 1967." *The Chinese Journal for the History of Science and Technology*, no. 4 (2002): 80–90.

Zhang, Huixuan. "The Cradle of Seamless Steel Tubes in China—Angang Seamless Steel Tube Plant." *Angang Technology*, no. 5 (1993): 56–59.

Zhang, Li. "5G Standard Ushers in the Key Point, China Leads Global 5G Development Competition." *China's Foreign Trade*, no. 7 (2018): 63–65.

Zhang, Shumin, and Xu Shushan. "Chen's Theorem and Its Application—A Theorem Named After A Chinese." *Physics Bulletin*, no. 5 (1994): 3–5.

Zhang, Zechun. "Six-Channel Neodymium Glass Laser Plasma Physics Experimental Device." *Mechanics in Engineering*, no. 2 (1981): 83.

Zhou, Yu. "Dongfanghon-I: Opening China's Space Age." *Science News*, no. 9 (2018): 22–25.

Zhu, Yehua. "Review of Hot Science and Technology Events in 2016." *Science and Technology Review*, no. 1 (2017): 138–150.

# Index

International Council of Science Unions (ICSU), 214, 215

Internet+, 290

## J

Jiang Zemin, 142, 179, 194, 213

Jiaolong manned submersible, 255

Jiuquan Satellite Launch Center, 91, 188, 189, 235, 251, 266, 283, 285, 287, 288

## K

Kunlong AG600, 277

## L

LAMOST, 211, 212

Lanzhou–Xinjiang Railway, 20

lasers, 86, 106, 116, 262, 311

leprosy, 84

Li Denghai, 110, 111

Lin Lanying, 31, 32

Li Siguang, x, 3, 4, 37, 66, 67, 175

locusts, 52, 53

Long March 3B launch vehicle, 251, 266, 267, 286, 288

Long March 5, 284, 285

Long March 6, 254

Long March 7, 284, 287

Long March 11, 285

Lunar Exploration Project (CLEP), 237

## M

maglev, 212

malaria, 85, 312

manned deep submergence, 250

Manned space program, 188, 235, 236, 251, 305

Mao Yisheng, 42, 67, 100, 242

Mao Zedong, 24, 26, 39, 48, 71, 91

material science, 114, 281, 309

Micius Satellite for Quantum Science Experiments, 285

microgravity science, 234, 235

mining, 8, 38, 113, 114, 124, 161, 171

## N

Nanchang CJ-5, 13

Nantong Machine Tool Factory, 32

Naraoia, 131

National High-Tech R&D Project (Project 863), 136

natural gas, 156, 207, 208, 209, 231, 290

Natural Science Foundation of China (NSFC), 136, 147, 150, 195, 309, 314

neutrino oscillation, 250, 257

nuclear fusion, 27, 86, 141, 167

## O

offshore oil Drilling Platform, 124, 126, 269

Olympic games, 200, 245

One Rocket & Three Satellites, 132

optical fiber, 168, 191, 212, 288

oxygen coal intensified ironmaking technology, 190

## P

Pacific Ocean, 74, 255

Paleoclimate, 89

Pan Jianwei, 261, 275

Panzhihua Iron and Steel Base, 77, 125

penicillin, 55

photon quantum computer, 275

polio vaccine, 55

Project 863, 136, 153, 171, 184, 211, 212, 229, 233, 255, 258, 265, 303

## Q

Qaidam Basin, 36, 157

Qian Xuesen, x, 9, 10, 28, 29, 95, 100

Qingdao Sifang Locomotive Factory, 14

Qinghai–Xizang Railway, 204, 298

Qinshan Nuclear Power Plant, 150, 151, 188

Qiu Shibang, 53

quantum communication, 250, 285

## R

Radio telescope, 270, 274

raw material technology, 162

real-time precision positioning technology, 233

Red Flag Canal, 76, 77

replantation, 56

rice, 52, 53, 75, 108, 109, 110, 150, 152, 153, 186, 204, 230, 262, 264, 265, 283, 303

robots, 194, 243, 276

Ruby laser, 34

## S

Sanjiang Plain, 108, 109

Xinyu Iron and Steel, 40, 41
Xue Qikun, 257
Xu Rigan, 122, 123

## Y
yeast, 129
Yellow River, 17, 20, 46, 73, 108, 209
Yinhe supercomputer, 131, 144
YTO Group Corporation, 280
Yuan Longping, x, 75, 110, 150, 265, 303

## Z
Zhang Wenyu, 7, 100
Zhang Yuzhe, 61, 100
Zhongguancun, 271, 272
Zhou Enlai, 24, 25, 26, 50, 69, 71, 72, 77, 85, 150
Zhou Peiyuan, 8, 68, 69
Zhu De, 71
Zhu Xi, 63
Zou Chenglu, 64, 65

## ABOUT THE AUTHOR

The writing group of *Glorious China*, with academician Liu Jiaqi serving as the science consultant, consists of more than ten experts and scholars in their respective fields, and popular science writers and editors, who jointly form the writing committee and the editorial committee.